Alternative Education for the 21st Century

Alternative Education for the 21st Century

Philosophies, Approaches, Visions

Edited by
Philip A. Woods
and
Glenys J. Woods

palgrave
macmillan

ALTERNATIVE EDUCATION FOR THE 21ST CENTURY
Copyright © Philip A. Woods and Glenys J. Woods, 2009.
All rights reserved.

First published in 2009 by
PALGRAVE MACMILLAN®
in the United States—a division of St. Martin's Press LLC,
175 Fifth Avenue, New York, NY 10010.

Where this book is distributed in the UK, Europe and the rest of the world, this is by Palgrave Macmillan, a division of Macmillan Publishers Limited, registered in England, company number 785998, of Houndmills, Basingstoke, Hampshire RG21 6XS.

Palgrave Macmillan is the global academic imprint of the above companies and has companies and representatives throughout the world.

Palgrave® and Macmillan® are registered trademarks in the United States, the United Kingdom, Europe and other countries.

ISBN-13: 978–0–230–60276–2

Library of Congress Cataloging-in-Publication Data is available from the Library of Congress.

A catalogue record of the book is available from the British Library.

Design by Newgen Imaging Systems (P) Ltd., Chennai, India.

First edition: January 2009

10 9 8 7 6 5 4 3 2 1

Printed in the United States of America.

Transferred to Digital Printing in 2010

This book is dedicated to the memory of Professor Bill Boyd, a loving friend, fine scholar and noble soul whose spirituality lives on through his legacy

Contents

List of Figures ix

List of Contributors xi

Introduction 1
Philip A. Woods and Glenys J. Woods

1 The K20 Model for Systemic Educational Change and
  Sustainability: Addressing Social Justice in Rural Schools
  and Implications for Educators in All Contexts 15
  *Mary John O'Hair, Leslie A. Williams, Scott Wilson, and
  Perri J. Applegate*

2 Democratic Schools in Latin America? Lessons Learned
  from the Experiences in Nicaragua and Brazil 31
  Silvina Gvirtz and Lucila Minvielle

3 The Touching Example of Summerhill School 49
  Ian Stronach and Heather Piper

4 Quaker Schools in England: Offering a Vision of an
  Alternative Society 65
  Helen Johnson

5 A Buddhist Approach to Alternative Schooling:
  The Dharma School, Brighton, UK 83
  Clive Erricker

6 Islamic Schools in North America and the Netherlands:
  Inhibiting or Enhancing Democratic Dispositions? 101
  Michael S. Merry and Geert Driessen

7 On their Way Somewhere: Integrated Bilingual
  Palestinian–Jewish Education in Israel 123
  Zvi Bekerman

8	"Alternative" Māori Education? Talking Back/Talking Through Hegemonic Sites of Power *Hine Waitere and Marian Court*	139
9	Starting with the Land: Toward Indigenous Thought in Canadian Education *Celia Haig-Brown and John Hodson*	167
10	Montessori and Embodied Education *Kevin Rathunde*	189
11	Education for Freedom: The Goal of Steiner/Waldorf Schools *Martin Ashley*	209
12	Pathways to Learning: Deepening Reflective Practice to Explore Democracy, Connectedness, and Spirituality *Philip A. Woods and Glenys J. Woods*	227

Bibliography	249
Index	275

Figures

1.1	IDEALS and 10 Practices for High Achieving Schools	18
1.2	Implementation of K20 Model	19
1.3	K20 Model of Systemic Change and Sustainability	20
1.4	K20 Embedded Professional Development for Teachers	29
8.1	He aha tēnei: What is This Thing Called Māori Education?	142
9.1	Medicine Wheel Showing the Four Aspects of the Self-in-relation	170
9.2	Toward Aboriginal Thought in Education, Six Directions (after Eber Hampton)	173
10.1	Student Experience in Montessori and Traditional Schools	204
12.1	Features of Bureau-enterprise and Developmentally Democratic Education	236

Contributors

Dr. Perri J. Applegate, University of Oklahoma, United States, is a research associate at the K20 Center for Educational and Community Renewal at the University of Oklahoma. Dr. Applegate's publications focus on democratic education in rural schools and communities. She has twenty-two years experience in the public schools as an English and journalism teacher and librarian. She holds B.S.Ed., M.L.I.S., and PhD, all from the University of Oklahoma.

Professor Martin Ashley, Edge Hill University, UK, is head of research in Education and director of the Centre for Learner Identity Studies at Edge Hill University, near Liverpool, UK. His principal research interests are in gender, generation, social class, and spiritual beliefs as dimensions of learner identity. He has researched extensively and published widely on the subject of boys and singing and also maintains an active portfolio of work on young people and environmental sustainability. During 2005 he worked with Philip and Glenys Woods on the DfES Steiner Schools in England project and maintains an interest in Steiner education through the Waldorf Researching Education Network (WREN).

Dr. Zvi Bekerman, Hebrew University, Israel, is an educational anthropologist and teaches at the School of Education and the Melton Center, the Hebrew University of Jerusalem. He is also Faculty member at the Mandel Leadership Institute in Jerusalem. His main interests are in the study of culture, ethnic, and national identity, including identity processes and negotiation during intercultural encounters and in informal learning contexts. He is widely published in these areas and is coeditor (with Seonaigh MacPherson) of *Diaspora, Indigenous and Minority Education*. He is author (with Claire McGlynn) of *Addressing Ethnic Conflict through Peace Education*, published by Palgrave Macmillan.

Dr. Marian Court, Massey University, New Zealand, teaches courses in the educational leadership programme at Massey University College of Education in Aotearoa New Zealand. Before this she taught in primary, intermediate, and secondary schools and worked briefly as an EEO reviewer with the New

Zealand Education Review Office. Her research and publications have drawn on poststructuralist, feminist, and Kaupapa Māori approaches in explorations of social justice issues in education. Her recent writing has focussed on the impact of new public management accountability regimes in education; intersections between gender and ethnicity in teachers' work and leadership; and innovatory approaches to sharing school leadership, including a study, with Hine Waitere, of a bicultural coprincipalship.

Dr. Geert Driessen, Radboud University Nijmegen, the Netherlands, received a teacher's degree before continuing on to study educational theory. His PhD thesis focused on the educational position of ethnic minority students in primary education. Currently he is a senior educational researcher at the ITS (Institute for Applied Social Sciences) of the Radboud University Nijmegen, the Netherlands. At ITS, he also served as a member of the Works Council, Head of the Department of Education, and member of the Management Team. His expertise lies in the field of education in relation to ethnicity/race, social milieu, and sex/gender. Among other things, he was involved in the large-scale secondary school cohort study VOCL (400 schools, 20,000 students) and in the large-scale primary school cohort study PRIMA (600 schools, 60,000 students). Both cohort studies were initiated to monitor the Dutch education system in general and to evaluate policies such as the Educational Priority Policy. He also performed policy evaluations with regard to Bilingual Education Programs and Early Childhood Education Programs. In addition, he has served as a project manager of dozens of research projects. His major research interests include education in relation to ethnicity/race, social milieu, and sex/gender; other themes are parental participation and engagement; school choice; religion and religious schools; Islamic schools; integration and segregation; participation and citizenship; preschool and early school education; bilingual education; dialects and regional languages; educational policy; compositional and peer group effects.

Dr. Clive Erricker is county inspector for Religious Education in Hampshire, UK. Previously he was Head of the School of Religion and Theology and Reader in the Study of Religions and Education at the University of Chichester. He is codirector of the Children and Worldviews Project, joint-editor of the *International Journal of Children's Spirituality*, visiting scholar of the Hong Kong Institute of Education, Research Fellow of the University of Winchester, and Pilot Examiner for the International Baccalaureate World Religions Curriculum. He has written many journal articles and books on religion and education, including a volume on Buddhism in the Hodder World Faiths Series, which is now in its 4th edition.

Professor Silvina Gvirtz, University of San Andres, Buenos Aries, Argentina, has a PhD in Education. She is the director of the Master in Education of

Universidad de San Andres and researcher of the CONICET (National Council for Scientific and Technical Research). In 2003 she was awarded the John Simon Guggenheim Fellowship for the project: "A comparison of models of school governance in Argentina, Brazil, and Nicaragua," and in 2004 she was awarded the 20th Anniversary Prize of the National Academy of Education of Argentina for her book, De la tragedia a la esperanza. *Hacia un sistema educativo justo, democrático y de calidad*. She has published fourteen books, the last one of which is Going to School in Latin America, (Westport: Greenwood Publishing Group, 2008). She has also published more than twenty articles in refereed journals of different countries, such as England, Germany, Portugal, Australia, Israel, Venezuela, and Belgium. She is Director of the Educational Series of Granica Publishing, and Director of the Yearbook of the Argentine Society of the History of Education.

Professor Celia Haig-Brown, University of York, Canada, is a EuroCanadian professor in the Faculty of Education at York University in Toronto, Canada. She teaches courses in research methodologies and community education. Her research focuses on Aboriginal education. Her most recent publication *With Good Intentions: EuroCanadian and Aboriginal Relations in Colonial Canada* (2006) is a series of historical essays coedited with David Nock. She is the author of three other books and numerous articles. Her current project, *Education as Regeneration: Processes of Decolonization,* funded by the Social Science and Humanities Research Council of Canada, explores the place of Indigenous thought in educational institutions.

John Hodson, Brock University, Canada, is the research officer in the Tecumseh Centre for Aboriginal Research and Education at Brock University. John is of Mohawk descent, Turtle Clan, and has worked in Aboriginal education at the college, university, and community level in Ontario for over fifteen years. In addition, John is a member of a number of circles and associations dedicated to Aboriginal teacher education/research, and has coauthored, published, and presented a number of articles related to the subject. John is currently pursuing a doctorate at York University and is a recipient of a Social Sciences and Humanities Research Council, doctoral award.

Dr. Helen Johnson, Kingston University, UK, is director of research and reader in the School of Education at Kingston University, London. In the context of a multicultural society, her work investigates faith/voluntary sector and alternative schooling; social and national identity formation and global citizenship education; and critical pedagogies within the scholarship of teaching and learning. She researches from an organizational studies perspective, particularly focused on values, culture, roles, identity, and managerial processes— and their impact on organizational and staff development within universities, schools, and the public services in general. Her most recent book is *Reflecting on Faith Schools* (London: Routledge). She is currently writing a book on educational research methods to be published by Sage, London, in 2009.

Dr. Michael S. Merry, University of Amsterdam, Netherlands, is professor of Philosophy of Education at the University of Amsterdam, the Netherlands. His philosophical and research interests include educational ethics, political philosophy, minority education, and alternative pedagogies. His published work in philosophy of education covers a wide range of topics, including cosmopolitanism, African-centered pedagogy, indoctrination, and giftedness. He has also published widely on Islamic schools and the challenges that Muslim minorities face in Western societies. He is the author of *Culture, Identity and Islamic Schooling: a philosophical approach* (Palgrave, 2007). Presently he is working on a volume that examines the interlocking notions of identity and citizenship.

Lucila Minvielle, University of San Andres, Buenos Aries, Argentina, is a PhD candidate in Education. She is assistant researcher at the Universidad de San Andres in Buenos Aires, Argentina. She worked on the Simon Guggenheim Fellowship Project "A comparison of models of school governance in Argentina, Brazil, and Nicaragua" with Silvina Gvirtz. She is currently working on the study of large-scale governance reform projects in Latin America.

Professor Mary John O'Hair, University of Oklahoma, United States, is vice provost for School and Community Partnerships at the University of Oklahoma and Professor of Educational Leadership. She is the founding director of OU's K20 Center for Educational and Community Renewal, a statewide, interdisciplinary research and development center that connects the University with over 500 schools and communities across Oklahoma. During her tenure, the K20 Center has innovated learning through democratic IDEALS (inquiry, discourse, equity, authenticity, leadership, and service) and a four-phased model for systemic change. Receiving over $40 million in external funding from public and private sources, K20 is impacting student learning and engagement throughout Oklahoma. She has published numerous articles and books on systemic innovation, organizational learning, and school-university-community partnerships. Her most recent book published by Open University Press is entitled *Network Learning for Educational Change* (2005; coedited with Wiel Veugelers).

Heather Piper, Manchester Metropolitan University, UK, is a senior research fellow at the Education and Social Research Institute at Manchester Metropolitan University, UK. She is a qualitative researcher, and her interests span a broad range of educational and social issues including ESRC sponsored research at Summerhill School. Her voice in research practice and academic writing is typified by a contrarian approach, a broad-based and eclectic intellectual territory in sociology, philosophy, social policy, and a sensitivity to interprofessional concerns informed by her own experiences outwith the academy.

Dr. Kevin Rathunde, University of Utah, United States, is an associate professor of Human Development and Family Studies in the Department of Family and Consumer Studies at the University of Utah. He received his PhD from the Committee on Human Development at the University of Chicago. His research adopts an experiential perspective and is focused on adolescent development in families and schools, the role of interest and flow in lifelong learning, and the importance of nature for education.

Prof Ian Stronach, Manchester Metropolitan University, UK, is research professor of Education, Institute of Education, Manchester Metropolitan University, and will be taking up a similar post at Liverpool John Moores University, in 2008. He was editor of the British Educational Research Journal for twelve years (1996–2007) and has published widely in international journals. His research interests are in postmodernist methodology and theory, as well as in areas such as evaluation, research epistemology, and professionalism. Appointed "expert witness" at the Summerhill Tribunal case (2000), he has since done ESRC-sponsored research there, including "inspecting the inspectors."

Hine Waitere, Massey University, New Zealand, is Māori of Ngāti Tūwharetoa, Ngāti Kahungunu, Tūhoe and Tainui descent. She has taught courses focused on bicultural and multicultural education in Te Uru Maraurau: The Department of Māori and Multicultural Education at Massey University, Aotearoa New Zealand. She is currently teaching in the educational leadership programme in the School of Educational Studies at Massey, writing from a mana wahine and kaupapa Māori perspective on issues of cultural leadership and equity in education. Hine's tertiary teaching and research interests emerge out of teaching experience in the primary sector, having taught in general stream, bilingual, rural, urban, and international schools. She was awarded a Fulbright Scholarship to study at the University of Wisconsin, Madison, and is currently completing her Phd from Wisconsin, Madison, with Michael Apple. She draws on the work of Bhabha and Bhaktin within a Kaupapa Maori framework as she explores issues of indigeneity and the ways in which the politics of knowledge, method, and praxis underpin a politics of healing in uneven worlds.

Dr. Leslie A. Williams, University of Oklahoma, United States, is a research associate professor in the Department of Educational Leadership and Policy Studies and Associate Director for Research at the K20 Center for Educational and Community Renewal at the University of Oklahoma. Dr. Williams has published numerous research articles and scholarly chapters on educational reform, learning community development, professional development, building leadership capacity to sustain systemic school improvement, and the integration of technology in teaching and learning processes to improve student achievement. She has been a Co-PI (Co-Primary Investigator) for over $27 million in grants supporting K12 school improvement in Oklahoma. Dr. Williams has been an educator for nearly twenty years, serving as a 6th–12th grade mathematics teacher, middle school principal, district mathematics curriculum director, and university professor.

Dr. Scott Wilson, University of Oklahoma, United States, has served as a middle school/high school science teacher, technology coordinator, coach, library media specialist, instructional materials coordinator, instructional technology coordinator, educational technology team leader, science curriculum coach, and a federal grant administrator in both a small rural and a large urban school settings. He has B.S. in Natural Science Education from Southwestern Oklahoma State University, MLIS in Library and Information Studies from the University of Oklahoma, and a PhD in Instructional Leadership and Academic Curriculum from the University of Oklahoma. His research interests include: digital game-based learning, technology integration, professional development (specifically technology integration PD), and management of constructivist classrooms.

Dr. Glenys J. Woods, University of Gloucestershire, UK, is research fellow in the Department of Education, University of Gloucestershire, UK. Her foundational interest is in spiritual awareness. She has undertaken research into spirituality, educational policy and leadership, and is coauthor of the first comprehensive review of Steiner schools in England. Current work includes research into alternative education and into the academy schools program in England, which is a continuation of research into the modernizing leadership agenda and its ethical challenges. She is also writing a book, for a wide readership including parents, on holistic and spiritual awareness and implications for balanced unfolding of spiritual, emotional, and cognitive development.

Professor Philip A. Woods, University of Gloucestershire, UK, holds a chair in Educational Leadership and Policy in the Department of Education, University of Gloucestershire, UK. He is author of *Democratic leadership in Education*, published in 2005, and has written and researched extensively on educational policy, leadership and governance, as well as exploring issues to do with creativity and values-intuitive social action in sociological theory. Current research interests include democracy and educational leadership, democratic learning communities, diversity and collaboration amongst schools, alternative forms of education, and entrepreneurialism and public values.

Introduction

Philip A. Woods and Glenys J. Woods

*An impossibly distant black bird
circled overhead and wondered why
so many bite-sized creatures spent their lifetime
running on the spot.*[1]

Deepening Reflective Practice

This book is a tool for deepening reflective activity—for educational practitioners, leaders, and policy makers in all types of educational settings—as well as a resource for academics and researchers. A key purpose of the book is to help educators think differently and creatively about education, and their own practice and/or policies, at a time when the need to address directly and imaginatively the fundamental human purpose of education, and how it can be realized for all in modern times, has never been greater.

The power of ideas that differ from those to which we have become accustomed is that they enable us to reflect anew on what is taken for granted. It helps what Bolton (2005, 85–86) calls "through-the-looking-glass" learning, in which the practitioner's familiar context is made as "strange and different" as possible and he or she is encouraged to be "as reflexively aware and questioning as possible of social, political, and psychological positions." Examination of alternatives is one of the ways to engage in "comparative reflection," to test and challenge "tightly held views [which] block us from considering . . . new ways of thinking and practice," and to open ourselves to "reframing,"—a "process of coming to see things from a different point of view or from within a new framework"—all aspects of reflective practice (York-Barr et al., 2006, 46).

The book is intended to be a resource for professional development and change in educational institutions—wherever practitioners and policy makers are looking for innovative ways of creating an ethos, curriculum, and pedagogies that are inclusive, open to alternatives, educate for and within democratic society, and nurture rounded physical, spiritual, emotional, cognitive, ethical, and social development in students. It encourages not only reflection on the internal practices and policies of the school, but also on external relationships and the value of the school actively cooperating with other (mainstream and alternative) educational settings in wider democratic learning communities. The book engages, from differing perspectives, with the profound purposes of education; and thus aims to *deepen reflective practice* and to stimulate ways of "creatively visualising the future" (Joy, 2006, 7).

We invite readers to utilize their knowledge and experience of their own context and to engage in *reflexive translation*[2]—by which educators engaging with the book use their knowledge and experience to consider if and how the alternative types of education discussed in the chapters are relevant to their own educational setting and community.

The context for this reflective practice is one characterized by continual change, globalization, modernization, privatization, school diversity, and attempts to enhance "consumer" choice. These are concepts that transcend national borders and characterize the policies of many different governments on public services, and they are integral to the neoliberal and third-way reforms that have reshaped and continue to work their way through school education (Ball, 2006). One of the most compelling critiques of this swathe of educational reform, which has swept so much before it, is that its understanding of child development and the purpose of education lacks depth. The political fervor to increase choice and to create new forms of schooling is too often beset by a focus on baselines and attainments that are measurable and an overriding concern with creating adults that will fit the perceived demands of the globalized economy. This framework of priorities is the environment that educators inhabit. And such an environment privileges certain goals, assumptions, and practices, whilst marginalizing others. It privileges the treatment of young people as potential units of economic activity, into which economically valued skills and attitudes (such as flexibility and team working) need to be inculcated, together with attributes that will sustain social cohesion (such as a sense of citizenship and civic responsibility). Considered as educational aims on their own, there is nothing wrong with such skills and attributes in themselves. However, the educational approach into which they are fitted and by which they are forged acts to restrict the educational vision of human development to one that sees people defined and valued, ultimately, as instruments of a given socioeconomic system. The predominant and effective aim—the purpose that becomes institutionalized into the system—is to shape people so that they *become* within their inner being the enterprising and instrumentally driven personality, subject to means–end rationality, which is prized by modern markets and new forms of bureaucratic organization (Courpasson and Clegg, 2006; Woods, 2001, 2007a).

Now, this is not to say that all educators are working to achieve this aim. The communities of educators in different countries are generally motivated and enthused by a commitment to develop the potential of children as young human beings. The question is what visions of alternative forms of education can feed into the deep reflective practice needed to progress this commitment in environments that tend to privilege a more restrictive view of education.

The Alternatives in This Book

The ideas in this book—better described as ideas-in-practice, as they are all educational philosophies that are being practiced in schools—are leading examples of forms of education grounded in alternative philosophies and cultures. The book purposefully juxtaposes different, distinctive, and contrasting alternative forms of education. These are not presented as perfect or near-perfect models—indeed their limitations are often explicitly acknowledged by the contributors to this volume—but as stimulants to reflective practice that can benefit from learning about both the successful and less successful features of alternatives.

We take a fairly pragmatic view for the purposes of this book of what constitutes mainstream education, thinking of it as the main conventions of publicly funded school education as generally understood in Western countries such as the UK, United States, Canada, Australia, and New Zealand. However, we recognize that mainstream education is itself not a monolithic entity, but varies in its form between and within countries and contains differences and alternative perspectives—sometimes representing the distinctive pedagogy or innovations of teachers or a particular school; sometimes being embedded more systemically, such as faith schools in the UK, which are part of the publicly funded system and have their own distinct ethos.

We also see that educators who are part of an alternative form of education may learn from the alternatives and discussion in this book and be encouraged to reflect in new ways on their own practices and approach. The stimulation and reflection that we hope this book will encourage is not just from alternatives to mainstream but in the opposite direction too, as well as between alternatives. Underpinning the book is a commitment to the idea that educators everywhere can benefit from seeing themselves as part of a large and diverse, international and intercultural democratic learning community. The promotion of both greater diversity in schooling and of collaboration and mutual professional learning between diverse school settings are strong trends in current educational policy, which offer both opportunities and challenges to educators (see, for example, Fielding et al., 2005, Woods et al., 2007, Woods and Woods, 2002).

In choosing the eleven alternative educational forms to be included in the book, we had a number of criteria in mind. The main aims underpinning the selection were to provide:

- a contrasting set of differing approaches that will stimulate professional reflection amongst educators and be an aid to innovative professional development,
- alternatives that are operating in practice as, or in, schools,
- an international range of alternatives,
- alternatives that are likely to be less familiar to mainstream educators,
- alternatives likely to evoke the interest of mainstream educators interested in developing progressive education.

Specifically, we wanted to include examples of non-faith alternatives concerned with developing democratic education (in the UK, United States, and Latin America); faith-based schooling on which relatively little is documented in Western contexts (such as Quaker and Buddhist schools) or which is likely to be of particular current interest (Islamic schools); examples of educational developments that concern intercultural education (Palestinian–Jewish schools) and indigenous cultures (Māori in Aotearoa New Zealand and First Nations in Canada); and schools that are based on educational philosophies (Montessori and Steiner/Waldorf) concerning which there is recent research that is illuminating and of general interest.

Developmental Democracy and Multiculturalism

Developing democracy in modern society is a complex and demanding process. For example, there are many diverse and sometimes conflicting conceptions of democracy, and different contending views of what democratic citizenship means have been evident in the United States, for instance, in response to the terrorist attacks of September 11, 2001, in New York and Washington (Westheimer and Kahne, 2003). Substantial disparities exist between different broad models of democracy. For example, the liberal minimalist model of democracy is a protective model, its main purpose and justification being protection of the individual citizen from arbitrary rule (Stokes, 2002). The assumed inherent capacity for human growth (socially, ethically, and spiritually) implicated in this model is narrow. On the other hand, a more expansive concept like that of *developmental democracy* aims to create an environment that enables the capacity for human potential—for intellectual reasoning, aesthetic sensibilities, spiritual awareness, and so on—to be realized; and it asserts that, as well as valuing participation in decision-making, consultations, and debates, this development of human potential is centrally important to democracy (Woods, G.J., 2005). Moreover, the inbuilt potential for each person to feel and understand what should be counted as true and of highest worth, founded in a human capacity for spiritual awareness (Woods, G.J., 2007), is at the heart of developmental democracy.

A concern with such an expansive concept of democracy, and with the deeper development of the person bound up with it, necessitates addressing the issue of diversity and cultural differences in modern society. If education is a contested concept (Boyd, 2000), and diverse cultural and other perspectives are to be respected in contemporary settings, how can it be possible to talk about a shared view of what schools can and should do in creating democracy? How can the deeper unfolding of young people's growth and identity be engaged and nurtured, without falling into the trap of indoctrination?

More and more national and ethnic communities within states have been seeking in various ways greater cultural recognition (Axtmann, 2001, 36–37), which makes it particularly important to understand the "dualities of democracy" (Woods, 2005, 131) that recognize the interconnectedness and the tensions between diverse cultural/personal identities and the unity of cohesive democratic society. One of those dualities recognizes the salience of both *substantive* and *protective* principles. Substantive principles emphasize the importance of a democratic order generating a sense of belonging and unity and of what it means to be a good person in a good society, while protective principles emphasize respect for diversity and rights to exercise freedom and to be different (thus challenging belongingness, unity, and a shared sense of the good).

A review of the international context for leadership emphasized the need for educational leaders to understand and engage with the diverse, pluralistic, and multicultural forces that characterize a globalized society and identified "deep democracy," which involves "respect for the worth and dignity of individuals and their cultural traditions," as "one of the most powerful emerging concepts" for educational leaders (Mulford, 2005, 150). One of the problems with much democratic theory, however, is that it comes with a "presumption of uniformity" (Hindess, 2001, 92)—an assumption that the citizens in a democracy are predominantly of one culture. There is an inbuilt tension between this presumption and multiculturalism, and hence democracy in a multicultural society needs to encompass a duality that balances substantive and protective principles. A healthy multicultural perspective involves the willingness and capability to work with this duality, to communicate across cultural boundaries, engage in critical reflection, and learn from differing traditions and viewpoints. Such a healthy multiculturalism expresses the spirit behind this book. Those who wish to learn from the opportunities offered by a multicultural perspective can

> set up a dialogue between [differing cultural vantage points], use each to illuminate the insights and expose the limitations of others, and create for [him- or herself] a vital in-between space . . . from which to arrive at a less culture-bound vision of human life and a radically critical perspective on [his or her] society. (Parekh, 2006, 339)

No single culture or philosophy or religion or tradition or view of life can, in this perspective, offer the one and true vision of the good life—since all have their limitations, blind spots, and capacity for error. However, as Parekh concedes, some shared preconditions are needed that provide a framework for such dialogue, which include

> freedom of expression, agreed procedures and basic ethical norms, participatory public spaces, equal rights, a responsive and popularly accountable structure of authority, and empowerment of citizens. (Parekh, 2006, 340)

Enlarging the Purpose of Education

The framework of shared preconditions proposed by Parekh involves an expansive conception of democracy. Any serious attempt to empower citizens is predicated on an understanding of people's potential powers and capabilities. In addition to capabilities that include critical and communicative skills, the sort of "deep democracy" that demands interpersonally and interculturally engaged, caring, and imaginative citizens, "calls for disciplined spiritual attunement and insight" (Henderson and Slattery, 2006, 2). Indeed, the relationship between the outer democratic order and the inner life of individuals is a dialectical one, each constituting the other in perpetual interaction (Woods, G. J. and Woods, P. A., 2008).

A developmental perspective on democracy, which takes an expansive view of the person, involves, therefore, enlarging the purpose of education beyond an instrumental focus. One of the most important foci for such enlargement is the school improvement and effectiveness agenda. Its perspective needs to be recast so that it focuses on education that fosters holistic development, enables young people to embrace and engage in democratic citizenship, and nurtures meaning (cultural and personal identity, moral and spiritual development)—which requires a movement in values: from predominance of instrumental rationality, efficiency, and compliance, to embrace empowerment, caring, community, connectivity, and inner development.

In particular, it involves, in our view, taking seriously the inherent spiritual capability of human beings—the potential for what Glenys Woods terms *quintessential spiritual experience*; that is, "a sense of being in touch with an inner strength which is also *part of,* or continuous with . . . a beneficent power far greater than the individual self" (Woods, G.J., 2007, 137).

The nurturing of *intelligent spirituality*, which in Alexander's (2003) conceptualization integrates the spiritual evocations of transcendent reality, inner reflection, and social solidarity, is a democratic imperative because it is the way to foster the virtues and sense of goodness necessary for democratic society. Hence, integral to our perspective on alternative education is a concern with the elevated aims of education that look to the higher meaning of human development—in other words, with the substance of education that is "hollowed out" by neoliberal and conservative reforms.

The Chapters

The first three chapters address differing examples of schools seeking to develop democratic education.

Chapter 1 addresses the development of democratic learning communities. Entitled "The K20 Model for Systemic Educational Change and Sustainability: Addressing Social Justice in Rural Schools and Implications for Educators in All Contexts," by Mary John O'Hair, Leslie A. Williams, Scott Wilson, and Perri J. Applegate, it focuses on a unique university-schools network to support the development of democratic schools. The chapter describes the Model for Systemic Educational Change and Sustainability, which has been developed by the K20 Center for Educational and Community Renewal at the University of Oklahoma in the United States. The K20 Model is dedicated to the creation of democratic learning communities by sharing best practices and cooperative learning, and, whilst it includes the components of a professional learning community, it emphasizes, in addition, equity inside and outside the school and the transformation of inequitable conditions. The Model comprises a detailed apparatus—such as frameworks of ideas and phases of implementation—that research and assessments suggest are working well. Factors that the experience of Mary John and her colleagues suggests are important to advancing in practice democratic schools include a robust vision of democratic learning communities grounded in research, the key role of a democratic leader in the school, and sustained, long-term support for schools.

In chapter 2, "Democratic Schools in Latin America? Lessons Learnt from the Experience of Nicaragua and Brazil," Silvina Gvirtz and Lucila

Minvielle draw from research undertaken into experiments in democratic education in Latin America. They report findings from a study that investigated attempts to forge alternative school education systems that facilitate democratic involvement and, through that, contribute to building up parents' and others' democratic capacities as citizens. Its findings show that in practice there is a tendency for nonprofessionals' participation to be "peripheral" or "passive," and for it to be inhibited by a dominant school culture that reinforces the idea that expertise rests exclusively with professionals. However, Silvina and Lucila also highlight features that appear to be more successful in encouraging and nurturing nonprofessionals' democratic participation, through less structured and more informal participation spaces. One of the strong messages from this research is that progress toward more democratic ways of working is a long process that requires deep change in school culture.

In chapter 3, "The Touching Example of Summerhill School," Ian Stronach and Heather Piper give a distinctive, research-based view of the famous, pioneering democratic school established in Britain in the 1920s. They undertook fieldwork as part of a wider study into interpersonal touching, but found that this cannot be understood sensibly without in-depth understanding of the community that Summerhill constitutes. The result is a perceptive account of the ways in which Summerhill's unique arrangements and culture create a climate in which, as Ian and Heather describe it, students are able to become themselves. Rather than Summerhill being about total freedom, the community is characterized by powerful, but invisible, boundaries—school pupils/"downtowners" (local inhabitants), locals/Summerhill cosmopolitans, limits to parental visits, and hidden definitions of normal/abnormal—as well as the central fixed point that is the decision-making body—the Meeting—to which all members of the Summerhill community (staff and students) belong. Within the context of these powerful boundaries and fixed spaces, the culture and Laws of the school community are cocreated by all its members (staff and students); and possibilities for change, creative expression, and resistance infuse the community through daily interpersonal negotiations of weaker organizational boundaries and spaces and through the Meeting. This makes Summerhill, as Ian and Heather describe it, both a strong and benign society, and, if not a perfect one, one that helps to nurture selves who are highly intuitive, are capable of negotiating tacit norms and reading other people, and have a real sense of belonging and who own the Meeting's judgments as *our* judgments.

Chapter 4, "Quaker Schools in England: Offering a Vision of an Alternative Society," by Helen Johnson, turns to a different kind of schooling, the first in this collection to address faith-based alternatives. However, the chapter can also be seen as offering insights into the distinctive expression of themes emerging from the previous chapters—for example, the theme, which appears in the chapter on Summerhill School, of *becoming* within a culture that expresses and encourages ways of living that place a prime value on democracy, mutual respect, and negotiated resolutions of conflicts and problems. Helen explores Quaker schools as sites where a particular culture can be experienced. Setting out the historical foundations of Quakerism, the chapter emphasizes its lack of dogma and its core features of providing opportunities for attentive silence, respect for the spirit of God (or the spirit of the divine) that is within every person, and commitment to peace. Helen concludes that the argument for the existence of Quaker schools is that they allow those sharing their environment to have experience of certain values, to live out another way, and to carry that experience with them for the rest of their lives. They nurture hope.

In chapter 5, "A Buddhist Approach to Alternative Schooling: The Dharma School, Brighton, UK," Clive Erricker provides a unique insight into a school that follows Buddhist teaching in a Western setting. A concern with developing such qualities as mindfulness and respect for living things is paramount in the school. Like the Quaker schools, the Dharma School and the Buddhist teachings in which it is grounded renounce fixed dogma and aim to create the conditions in which people can live together peacefully and develop the capabilities to deal with conflict and live in awareness and mindfulness—of themselves, others, and nature. Thus themes of becoming and conflict resolution appear here too. The chapter places the school, its vision, and pedagogy in the context of the Enlightenment value of preparedness to critique all knowledge claims and the importance of developing critical capabilities for democracy. However, far from requiring unquestioning faith, Buddhist tenets, Clive argues, ask those who aspire to follow the Buddhist path not to be "idolatrous about or bound to any doctrine, theory or ideology" (Carey, 1992, 21). Key to understanding the Dharma School is, therefore, that it takes the critical spirit of the Enlightenment and combines this with virtue, compassion, and spirituality, thus restoring to education a concern with "virtue ethics." In this way, it aims to nurture persons who fulfill their full potential and grow into rounded citizens of democratic society.

The next chapter turns to a different kind of faith school, serving a faith community that has grown considerably in Western countries. Michael S. Merry and Geert Driessen in chapter 6, "Islamic Schools in North America and the Netherlands: Inhibiting or Enhancing Democratic Dispositions?," examine Islamic schools in those countries—that is, what they describe as mainstream, full-time Islamic schools that endeavor to provide a state-recognized education in all subject matter necessary for ordinary primary- and secondary-level studies. They assess the strengths and weaknesses of Islamic schools, drawing on their own considerable research experience in this field, as well as citing other studies and evidence, and thereby make a valuable contribution to understanding these schools and to raising important issues concerning their future development. There is a great deal of variation amongst Islamic schools—in their contexts, resources, and so on—and in both their academic and nonacademic outcomes. So it is perhaps not unsurprising that their conclusion on whether Islamic schools are sites for democratic learning is mixed: some are, others are less likely to be so. This, the chapter concludes, is a major challenge for Islamic schools—if they exchange one type of conformity (peer pressure in public schools) with another of their own (unquestioning acceptance of rules or gendered expectations), there is reason to ask, Michael and Geert argue, whether they are meeting their own standards of excellence and equality. At the same time, examples of openness to critical reflection are cited, such as the American Islamic high school where many students are confident to question the principles of Islam or the authority of their teachers.

The three chapters that follow are, in different ways, about how cultures come together or are sustained, nurtured, and developed through school education, and the struggles in creating counter-hegemonic educational spaces for peoples and cultures that are disadvantaged. In chapter 7, "On their Way Somewhere: Integrated Bilingual Palestinian Jewish Education in Israel," Zvi Bekerman gives an unparalleled insight into the progress and challenges in schools that seek to bring together Palestinian and Jewish communities. A unique initiative in Israel, founded by a Palestinian and a Jew who saw this change as necessary for a genuinely democratic society, has given rise to four integrated, bilingual schools to which Jewish and Palestinian parents send their children, the first of which opened in 1997 and the most recent one in 2007. Zvi has researched the schools over an eight-year period and demonstrates that this is a project still very much in development, with examples of great successes (such as the creation and maintenance of a well-balanced Palestinian and Jewish staff and finding ways to recognize the respective commemorative days of the two communities) and lesser successes (such as the attempts to maintain a balance between the Arabic and Hebrew languages). The power of the sociopolitical context and its power relations cannot be overcome simply by the creation of such schools. But, whilst challenging for both Palestinian and Jewish parents in different ways, they offer signs of progress, including, for Palestinian parents, the promise of opportunities for their children to gain higher education and career chances otherwise not open to them. Whilst emphasizing that education cannot by itself change the social structures that underpin unequal relationships, Zvi also sees that this small initiative might contribute to a change of rhetoric that will inspire others to follow.

INTRODUCTION / 11

In chapter 8, "'Alternative' Māori Education? Talking Back/Talking Through Hegemonic Sites of Power," Hine Waitere and Marian Court ask challenging questions about the concept of "alternative" and explore in detail the struggles of Māori education in Aotearoa New Zealand, illuminating these through an appreciation of the Ngā Manu Kōrero (Māori speech competitions). Poetry that they have written for the chapter, examples of speech competition awards that illustrate the symbols and historical awareness of Māori peoples, and historical policy description and analysis provide insights into both the injustices suffered by Māori peoples and their education, and the positive action that Māori peoples themselves have and are taking to sustain and enhance their identity through education. The conceptual framework Hine and Marian posit and use in their exploration sets out three types or meanings of alternative: *alternative*, meaning Māori education as just another option within or peripheral to dominant mainstream education; *alter-native*, representing the use of Māori education as an assimilatory strategy, to "alter the native"; *(alter)native*, representing Maori education as the norm, or a norm, centered in and central to the mainstream, alongside Pākeha,[3] state education provision. In contemporary Aotearoa New Zealand, educational policy has moved toward a bicultural frame of reference that helps Māori education to be seen as "*(alter)native* mainstream." This, argue Hine and Marian, is not only to the benefit of Māori peoples but also the dominant Pākeha, as otherwise, for example, they are denied the opportunity to benefit from the knowledge, perspectives, and frames of reference that can be gained from studying and experiencing other cultures and groups. Achieving *(alter)native* status involves agency by Māori, which ensures that self-determination is not a gesture of charity or provision of an *alternative* option but an act of justice that realizes a democratic ideal that centrally locates Māori within the mainstream.

Chapter 9, "Starting with the Land: Toward Indigenous Thought in Canadian Education," by Celia Haig-Brown and John Hodson, explores questions concerning First Nations education in Canada: What happens when Indigenous ways of knowing and being in the world come to bear on Eurocentric forms of education and schooling? What are the possibilities for transformation of understandings, for shifts in world view, with deliberate efforts to interrupt classroom epistemological business as usual? Some responses to these questions are explored using "circlework teachings." Celia and John indicate how the circle is central to most Indigenous cultures: it refuses linearity and compartmentalization, emphasizing interwoven connections between and amongst things. Seeking regeneration and healing following the long centuries of colonization, First Nations education starts with awareness of the relationship with the land, and places the spiritual connections and fundamental respect for each other and for the earth at its centre. Celia and John emphasize that Indigenous thought should be understood on its own terms, starting from a complex engagement with the four interrelated physical, emotional, intellectual, and spiritual realities (the self-in-relation). Many of the diverse ways in which the Indigenous peoples in Canada are regenerating their esteem, self-awareness, and educational traditions are described. One example featured is "learning and healing" pedagogy, the contemporary expression of traditional Indigenous forms of education, which is a wellness model that understands that all are engaged in a healing journey and aims to promote and maintain balance in the individual, the family, the community, and the Nation.

The next two chapters consider Montessori and Steiner/Waldorf education. In chapter 10, "Montessori and Embodied Education," Kevin Rathunde focuses on Montessori schools. The first part of the chapter suggests why alternative visions of education are needed, arguing that many traditional schools have drifted toward *disembodied education* where students are drilled cognitively and suffer the experiential consequences of drudgery, lack of motivation, and mental fatigue. The chapter then goes on to articulate the distinctive features of Montessori education and suggests how they contribute to holistic human development, highlighting the close parallels with optimal experience (flow) theory (Rathunde and Csikszentmihalyi, 2006). The final part of the chapter discusses recently completed research supporting the claim that Montessori education is associated with student engagement and flow. This study provided some confirmation that the embodied/experiential focus of Montessori education pays off in terms of student experience.

Chapter 11 turns to Steiner/Waldorf education, which is based on the philosophy (anthroposophy) of Rudolf Steiner. Its curriculum and pedagogy aim to awaken the range of human faculties, with the aim of encouraging balanced growth toward "physical, behavioural, emotional, cognitive, social and spiritual maturation" (Rawson and Richter, 2000, 7):

> At its heart is a spiritual process. As these faculties open up, "the spiritual core of the person [strives] to come ever more fully to expression" (Rawson and Richter, 2000, 7). And the purpose is for each child to come to their own individual expression. The stated aim of Steiner education is not to create adherents of anthroposophy. Rather, it is to awaken young people to the spiritual and ethical dimensions of human life and to enable them to be free and independent thinkers and to make decisions for themselves. (Woods and Woods, 2006c, 317)

In chapter 11, "Education for Freedom: The Goal of Steiner/Waldorf Schools," Martin Ashley explains that at the very core of anthroposophy lies the deep-rooted philosophy of freedom that seeks to liberate the human condition through the integration of science, art, and religion. The chapter also draws on research that Martin and the two of us (Woods, P. A. et al., 2005) carried out into Steiner/Waldorf schools in England. A particular strength of Steiner/Waldorf education that Martin highlights is the stability for children offered through the sustained relationship over several years between a class and its class teacher. This is consistent with the idea that spiritual growth and holistic development through close and caring relationships are a positive characteristic of the Steiner/Waldorf schools. Another feature highlighted is the two-hour "main lesson," which offers a well-refined system of curriculum integration that incorporates many good principles of "rhythm" or pace and timing. These provide, Martin argues, a better alternative than the fragmentation of both curriculum and pastoral support that often characterizes state practice.

In the final chapter (chapter 12), "Pathways to Learning: Deepening Reflective Practice to Explore Democracy, Connectedness, and Spirituality," we consider what can be drawn from the alternatives. The preeminent theme we identify is connectedness. The chapter briefly considers alternatives from the perspective of their differing positionings, power, and meanings, and argues that diversity is a valuable feature of democracy. It then sets out a framework of features that comprise education with a developmentally democratic character that aims to nurture the spiritual and aesthetic sensitivities, the secure identity and self-esteem, and the cognitive abilities and skills necessary to nurture independent-minded individuals with a sense of civic and ecological engagement. The themes and concepts in the framework represent our suggestions of what can be drawn from a consideration of the alternatives. Suggestions for reflective activity are included. These are pointers to areas of exploration that alternative and mainstream educators may follow in their own way, selecting from and utilizing, critiquing, and adapting concepts, practices, and experiences of the alternatives in previous chapters.

Notes

1. *Standing Stone Poem*, Paul McCartney, 1997. See CD, *Paul McCartney's Standing Stone*, published by MPL Communications Ltd. under exclusive licence to EMI Records Ltd., 1997.
2. This is similar to what Stake (1978) calls naturalistic generalization, in which researchers and practitioners assess the extent to which findings in one context apply (or are *transferable*) to other contexts that they know.
3. Non-Māori, generally of European descent.

CHAPTER 1

THE K20 MODEL FOR SYSTEMIC EDUCATIONAL CHANGE AND SUSTAINABILITY: ADDRESSING SOCIAL JUSTICE IN RURAL SCHOOLS AND IMPLICATIONS FOR EDUCATORS IN ALL CONTEXTS

*Mary John O'Hair, Leslie A. Williams,
Scott Wilson, and Perri J. Applegate*

Introduction

Forty percent of U.S. school districts, serving nearly 10 million children, are located in rural settings (Johnson and Strange, 2007). Of the 250 poorest communities in the United States, 244 are rural (Malhoit, 2005). One-fifth of all rural children in the United States live in poverty, and minority children living in rural areas are even more likely to be poor, with 46 percent of rural African American children and 43 percent of rural Native American children falling into this category (USDA, 2004).

Rural schools are at a disadvantage when competing for resources for professional development and attracting qualified teachers, with one in four rural science teachers lacking in academic preparation or certification (National Science Board (NSB), 2006). According to the National Education Association (NEA), 41 percent of teachers in the United States are employed by rural schools (NEA, 2003). Compared to their non-rural counterparts, rural teachers average 13.4 percent less in salary; live in substandard housing; experience professional, cultural, and social isolation; and receive little if any professional development (Beeson and Strange, 2003; Darling-Hammond, 2000; Education Trust, 2003; Jimerson, 2003). Rural principals and superintendents feel ill-prepared for the challenges that face them (Lamkin, 2006). Thus, although social justice is often discussed in terms of race, class, gender, disability, and sexual orientation, it may also be an issue of location—in this case, being located in a rural area (Applegate, 2008; O'Hair and Reitzug, 2006).

The primary purpose of this chapter is to examine an alternative for a neglected dimension of social justice in the United States—social justice for rural schools and, particularly, for the education of the students who attend these schools and the professional development of the educators who serve in them. We do this by describing the kindergarten through graduate education (K20) Model for Systemic Educational Change and Sustainability (O'Hair et al., 2005). This model was developed by the K20 Center for Educational and Community Renewal (K20 Center) at the University of Oklahoma. In the following sections, we review this alternative model in regard to (1) background, (2) a detailed explanation of the model and its theoretical basis, (3) two key defining features, (4) strengths and challenges, (5) implications for practice, and (6) additional sources.

Background

A decade ago, the University of Oklahoma (OU) made a novel investment designed to improve student learning by connecting K-12 schools across Oklahoma with OU faculty and students. The Oklahoma Networks for Excellence in Education (ONE) was formed in 1995 to create communities for democratic education. This network began with six elementary schools from urban, suburban, and rural settings and evolved to include secondary schools. The vision focused on creating democratic learning communities by sharing best practices and learning from each other. The democratic learning community, as defined by the K20 Center, includes the components of a professional learning community, but, in addition to those components, it emphasizes equity inside and outside the school and transforming inequitable conditions in both settings (Reitzug and O'Hair, 2002). In this way, democratic education has a transformative nature (Reitzug and O'Hair, 2002). Democratic education is "schooling *for* democracy and schools *as* democracies" (O'Hair et al., 2000).

Today, OU's initial investment of the K20 Center has paid off through the creation of a statewide educational research and development center that connects the University with over 500 schools. The K20 Center's research is focused on innovations for teaching and learning. Knowledge gained from these research activities is transferred to Oklahoma schools through the K20 Center's professional development for K-12 teachers, principals, and superintendents. The K20 Center targets low-income, rural schools serving diverse populations in Oklahoma (including the 22,000 Native Americans who attend rural Oklahoma schools). To date, the K20 Center has impacted approximately 500 schools, mainly rural, and more than 90,000 students. It helps reduce the inequitable conditions of professional, cultural, and social isolation and the lack of professional development in rural schools.

The K20 Model: Overview and Theoretical Basis

Findings from research on the collaboration among the initial network schools were the basis for the K20 Center's *IDEALS* framework (figure 1.1). *IDEALS* is an acronym representing Inquiry, Discourse, Equity, Authenticity, Leadership, and Service as key democratic principles. Inquiry is critical to a democratic learning community because it involves the study of a school's practice through gathering and considering data, new knowledge, and others' perspectives. Discourse refers to conversations, discussions, and debates focused on teaching and learning issues that nurture professional growth, build relationships, and result in more informed practice and improved student achievement. Equity, which refers to seeking fair and just practices both within and outside the school, is one of the key principles for a democratic learning community. Authenticity focuses on learning that is genuine and connected rather than something that is fake and fragmented. Teachers who practice authenticity help students connect learning to life and their community. Leadership in democratic learning communities involves the development of shared understandings that lead to a common direction and improves the school experience for all members. Democratic leadership is shared, empowering all stakeholders. Service is the final principle of a democratic community and refers to the belief that making a difference in the lives of children and families requires serving the needs of the community as well as the school. Through continued research of developing rich democratic learning communities, the K20 Center identified *Ten Practices of High-Achieving Schools* that lead to increased student achievement (figure 1.1) (O'Hair et al., 2005).

Scholarly investigation of network schools, along with external research, serves as the basis for the K20 Center's systemic school reform model. Findings from work with the initial network schools demonstrate the importance of a strong leader who supports and promotes the creation of a technology-enriched democratic learning community. Accordingly, the first phase of the systemic model targets school principals and superintendents. Following completion of this first phase, administrators are offered the opportunity to compete for funds to implement the next phase of the Center's model, whole-school reform. The third phase deepens this whole-school learning by targeting teachers, particularly those in science, technology, engineering, and mathematics (STEM). The fourth phase utilizes

Figure 1.1 IDEALS and 10 Practices for High Achieving Schools

innovative technology to engage today's students in authentic learning experiences. Figure 1.2 provides additional information on each phase of the K20 Center's systemic school reform model.

All four phases of the K20 Center's systemic model have been based on research that shows educator learning is greater when professional development utilizes an embedded approach, linked directly to student achievement (for example, lesson study, authentic experiences, democratic learning communities, advanced learning technologies), rather than the traditional workshop or conference format (Atkinson, 2005; Garet et al., 2001). Similarly, Fullan (2001a, 2003) notes that to significantly improve student learning, teachers also must be continuously learning. Based on this knowledge, the K20 Center developed an embedded professional development model that provides interactive instruction and authentic research

> **Year 1 – Leaders Learning:** The K20 Model for Systemic Change and Sustainability begins with school leader learning (superintendents, principals, technology directors, and other administration staff). The professional development focuses on developing strategies for transforming traditional schools into democratic learning communities by implementing the *IDEALS* and *10 Practices of High Achieving Schools*. Upon completion, leaders develop action plans that maximize the democratic *IDEALS* and *one of the 10 Practices* (i.e., shared vision for teaching and learning) while integrating technology into learning.
>
> **Year 2 – Whole School Learning:** Building on the initial action plan by a leader, a school selects three of the *10 Practices* to focus on whole-school learning. The K20 Center staff supports whole-school reform by providing professional development to build democratic learning communities using technology as a catalyst for changes in instructional practices. Expanding leadership capacity in the school through the creation of learning teams provides infrastructure for continued growth and sustainability.
>
> **Year 3 – Teacher Learning:** With supportive leaders and active learning teams in place, K20 emphasizes additional *10 Practices* to accelerate and support change from didactic to interactive pedagogy. Specifically, K20 (a) deepens teacher content knowledge and comfort with interactive pedagogy and technology; (b) helps transfer teachers' new learning into daily classroom practices; (c) connects teachers with higher education and industry mentors (i.e., scientists, artists, engineers, mathematicians) to enhance and build authentic learning experiences; and (d) creates democratic learning communities to provide continuous support for interactive pedagogy, such as regional networking and lesson study opportunities.
>
> **Year 4 – Interactive Student Engagement:** As teachers transfer their learning experiences to daily practice, technology-enriched interactive learning environments emerge. Virtual and service learning experiences help ensure all students have access and opportunities to interactive learning and to a culture of civic engagement beyond the classroom.

Figure 1.2 Implementation of K20 Model

experiences coupled with advanced learning technologies for rural teachers. Additionally, the K20 Model provides structures and opportunities for rural teachers to translate new knowledge and experiences into their instruction through lesson study, digital game-based learning, and service learning. For the past twelve years, the K20 Center has promoted systemic "whole school" reform through a school–university partnership designed to transform conventional schools into democratic learning communities. This transformation uses peer coaching, regional networking, technology transfer, and the *IDEALS* and *10 Practices of High-Achieving Schools* (Cate and O'Hair, 2007; O'Hair et al., 2000; O'Hair et al., 2005) to accelerate student and educator learning.

The K20 Center's efforts are grounded in the K20 Model for Systemic Educational Change and Sustainability illustrated by a continuous generation of deeper understanding and application of the *IDEALS* and *10 Practices* (O'Hair et al., 2005). The K20 Model is characterized by the ongoing relationship between the K20 Center and K-12 network as participating schools take risks to reform themselves from traditional schooling practices to democratic communities that are characterized by innovation, creativity, and student engagement.

K20 professional development moves beyond a conceptual framework to one that is evidence-based and promotes exemplary instructional practices in rural classrooms. Transforming a school culture is a highly complex, multivariate process (Cate et al., 2006) that takes from three to five years (Fullan, 1991). By integrating constructive approaches and systemic change that begins with school leaders and moves eventually to every teacher, student, and parent, the K20 Model creates and helps sustain technology-enriched learning communities across the state. Figure 1.3 depicts the change process involved in implementation of the K20 Model for Systemic Change and Sustainability.

The K20 Model responds to the contextual needs of each school and, in the process, documents progress and outcomes. In most schools, the K20 Center has documented a period of "generalized skepticism" that ultimately succumbs to a "tipping point" where productivity and creative implementation of the school's evolving vision ensues. This eventual success, coupled

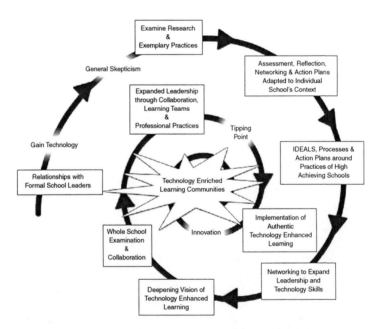

Figure 1.3 K20 Model of Systemic Change and Sustainability

with the increasingly effective infrastructure of the schools' learning teams, regional networking, technology integration, and service learning, spawns new creative and innovative educational goals and mechanisms for increasing and sustaining student success and overall school performance.

Two Key Defining Features

The K20 Model to promote the creation of democratic learning communities includes two defining features or components: authentic interactive instruction and job-embedded professional development. Each is supported by professional literature and practice. The concept of a democratic learning community is not new to education. From Locke's seventeenth-century work to that of Dewey in the twentieth, scholars have averred that schooling and democracy are inextricably connected. It is obvious to many that public schools and democratic ideals should be linked (Apple and Beane, 1995; Conant, 1959; Cuban, 2000; Darling-Hammond, 1997; Dewey, 1916; Fullan, 2001b; Furman and Shields, 2003; Glickman, 1993, 1998; Glickman and Aldridge, 2003; Gutman, 1993; Meier, 2000, 2002; Newmann, 1993; Noddings, 1996; O'Hair et al., 2000; Reitzug and O'Hair, 2002; Rusch, 1995; Starratt, 2003; Wood, 2005; Woods, 2005). Despite this long history, democratic learning communities are rare in the United States (Rusch, 1995).

Authentic Interactive Instruction

Authentic teaching, learning, and assessment consist of engaging students in "personal construction of new knowledge" and "disciplined inquiry" concerning work that has "value beyond school" (O'Hair et al., 2005, 75). Construction of knowledge "involves organizing, interpreting, evaluating, or synthesizing prior knowledge to solve new problems" (Newmann et al., 2001, 14). Disciplined inquiry requires students to use knowledge and strive for in-depth understanding, and then communicate this knowledge to others (Newmann et al., 2001). Value beyond school may be "utilitarian, aesthetic, or personal value" (Newmann et al., 2001, 15). Authentic learning helps to overcome the problem found in many classrooms of students being expected to be passive learners (Fouts and Associates, 2001; McLaughlin and Talbert, 2001; Meier, 2002; National Research Council, 2004; Shields, 2002a; Wood, 2005).

Many researchers and scholars consider authentic interactive instruction to be a key component for successfully reforming schools (Darling-Hammond, 1997; Fouts and Associates, 2001; Furman and Shields, 2003; Huffman and Hipp, 2003; Lambert, 1998; McLaughlin and Talbert,

2001; Meier, 2002; Murphy et al., 2001; National Research Council, 2004; Newmann, 1993; Newman et al., 2001; O'Hair et al., 2000; Reitzug and O'Hair, 2002; Sizer, 2004; Starratt, 2004; Wood, 2005). Newmann et al. (2001) make the point that authentic learning does not preclude students learning basic skills. "Students must learn basic skills, that is, essential knowledge, procedures, and conventions in writing, speaking, and computing to participate successfully in contemporary society" (Newmann et al., 2001, 13). Authentic interactive learning gives students the essential democratic skills of questioning and making decisions based on fact (Apple and Beane, 1995). It gives students the skills to become lifelong learners (Barth, 2005). Finally, authentic interactive instruction puts the responsibility for learning on the student rather than the teacher (Newell, 2005; Starratt, 2004).

Research suggests that meaningful learning for students:

- is active (Bransford et al., 2004);
- allows for student's choice (Yair, 2000);
- involves the acquisition of organized knowledge structures and social interaction (Piaget, 1972; Vygotsky, 1978; Greeno, 1997);
- relates new information to existing cognitive structures (Blumenfeld et al., 1991; Good and Brophy, 2000; Hannafin and Land, 1997; Jonassen, 1999);
- facilitates the development of knowledge that can be applied to different contexts and problems (Blumenfeld et al., 1991; Bransford et al., 2000);
- creates an intellectual climate that nurtures and challenges students and results in high student engagement (Yair, 2000);
- uses inquiry and other techniques that explicitly help students make sense of the content at hand (Newmann, 2003; Pasley et al., 2004).

Authentic interactive instruction is essential for democratic learning. Students develop necessary skills to function as productive citizens within a democracy. Authentic interactive instruction increases achievement for *all* students, regardless of school level, size, context, ethnicity, or socioeconomic status (SES) (Smith, Lee, and Newmann, 2001). This type of teaching results in students producing intellectual artifacts that are worthwhile, significant, and meaningful, such as those produced by successful adults (scientists and other professionals) who apply basic skills and knowledge to complex problems (Newmann, 1996; Newmann and Wehlage, 1995). Research supports the proposition that interactive instruction producing more authentic intellectual work improves student scores on conventional tests (Newmann et al., 2001) and student motivation to learn (Greene et al., 2004; Roeser et al., 1996). Concerns about the reluctance of teachers to implement authentic interactive instruction are not new (Blumenfeld et al., 1991). For students to receive interactive instruction that engages them in authentic intellectual work, teachers must learn new teaching methods and acquire more knowledge on subject matter. Newmann and Associates (2001) emphasized that asking teachers to effectively implement authentic intellectual work necessitates providing resources for integration and assessment and professional networking opportunities. In the K20 Center's systemic school reform model, job-embedded professional development addresses these needs.

Job-Embedded Professional Development

The K20 job-embedded professional development (PD) includes processes such as inquiry, discussion, evaluation, consultation, collaboration, and problem solving and is stimulated by new roles for teachers (for example, teacher leader, peer coach, mentor, teacher researcher), new school structures (for example, problem-solving groups, decision-making teams, common planning periods, self-contained teams), or new tasks (for example, leading an in-house workshop, journal writing, collaborative case analysis, grant writing, curriculum writing, school improvement team membership) (O'Hair et al., 2005). The K20 PD, described next, combines five research-based strategies linked to accelerating and supporting change in rural classrooms from didactic to interactive pedagogy. These PD strategies include:

1. *Authentic research experiences for teachers* to deepen understanding of inquiry while enriching substance and process (Kincheloe, 1991; Newmann, 2000).
2. *Lesson study*, credited with bringing about Japan's evolution in mathematics and science teaching (NRC, 2002); this is concerned with translating learning that occurs during teachers' research experiences directly into classroom practice.
3. *Digital game-based learning* (DGBL) to construct an authentic learning environment that allows students to explore inquiry-based, real-world activities in a virtual setting (Becker, 2007). This DGBL environment improves student motivation, increases engagement in learning (Ravenscroft and McAlister, 2006), and encourages critical thinking (Inal and Cagitay, 2007; Lim et al., 2006; Mitchell and Savill-Smith, 2004).

4. *Democratic learning community*, including peer coaching, regional networking, and technology integration, to support authentic interactive instruction. These learning communities have been associated with higher levels of student achievement (Atkinson, 2005; Lee and Smith, 1996; Newmann et al., 2001; Williams, 2006) and reduction of the remoteness and isolation that affects rural teachers' learning (Malhoit and Gottoni, 2003).

5. *Service learning*, which involves students giving back to their local community and becoming involved in community affairs, improves student academic learning (Dávila and Mora, 2007) and contributes to continued student involvement in community service after graduation (Astin and Sax, 1998).

Further details of the components, purpose, and key constructs of K20-embedded PD are described in figure 1.4.

Strengths and Challenges

The K20 Center models democratic principles of inquiry and discourse by continually reflecting on the results of its work through ongoing research, reflection with critical friends, and external evaluation of programs and projects. Research and assessments, which we cite in this section, provide evidence that implementation of the K20 Model improves rural teacher quality and student success while ensuring sustainability beyond K20's involvement (Hamlin, 2007). K20 partner schools report significant increases in high-achieving school practices of shared vision and leadership, collective learning, peer observation and critique, and technology access and transfer (SEDL, 2004; Williams, 2006). As K20 schools incorporate characteristics of democratic learning communities, a commensurate increase of technology integration into their practices, including use by students for learning, is documented (Atkinson, 2005; Williams, 2006). Teachers create new knowledge for technology integration and collaborate within and among classrooms for teaching and learning. Evidence from general research literature suggests that this kind of organizational learning with continual renewal increases the development of democratic learning communities and school capacity for change (Dufour et al., 2002; Hord, 1997; Huffman and Hipp, 2003; Sergiovanni, 1994).

Over the past four years, the State of Oklahoma's barometer for academic success (Academic Performance Index (API)) has recorded marked improvement in K20 partner schools (74 percent higher than the state's average increase) (Williams et al., 2007). Although it is certainly possible that other factors impact the API gains, the only attribute shared by all these schools is their involvement in the K20 Center's Systemic Reform Program. These results may well extend into higher education and industry as students complete a multitude of degrees, particularly STEM-related, and enter the workforce. K20 efforts provide a strong foundation to grow and advance cutting-edge research and development programs designed to respond to "America's urgent challenge to remain a world leader in science and technology" (NSB, 2006, 1).

Through the K20 Center's school–university partnership, a statewide network has developed to improve opportunities for school leaders, teachers, students, and communities in Oklahoma. The K20 Model supports mutual learning among these stakeholders while constantly expanding the learning partnership to include scientists, engineers, and community/industry/government leaders. The National Staff Development Council (2004) ranked the K20 Center's leadership program third nationally out of fifty state programs in developing educational leaders for systemic, substantive changes impacting student learning.

A challenge for the K20 Center remains building and supporting capacity for change in over 500 rural schools and communities with limited resources, staff, and time—the same barriers that most organizations face. Currently, K20 has a staff of over forty employees consisting of former principals, professional development and curriculum directors, teachers, as well as technology specialists and a multitude of university faculty and students who collaborate with K20 staff on various projects to enhance interactive learning and student success. With limited resources, K20 is unable to undertake in-depth work with all participating schools, but, rather, strives through intentional networking and job-embedded professional development structures to help schools learn from each other and colleagues in higher education and industry. K20 seeks to maximize its resources by enhancing the school's capacity to support and sustain its renewal work. Reflecting on K20 Center's growth during its twelve-year tenure, the resources challenge has diminished from the early years as K20 has established its ability to identify, analyze, and significantly address educational issues at the local and state levels.

A further limitation is based on the diverse funding that has made various phases of the K20 Model possible. Because different sponsors have specific evaluation requirements, funding cycles, and project priorities, research that evaluates the entire model has been limited. The programs that are involved in the K20 Model, with a multimillion dollar annual budget, are funded by diverse agencies such as the U.S. Department of Education, the National Science Foundation, and various state and local organizations. It would be advantageous to track the development of a democratic community through longitudinal analysis including each phase. As mentioned earlier, time is always a challenge, but longitudinal research is a high priority as the Center continuously looks to add to the literature-base on democratic schooling.

A final limitation is that the K20 Model has only been implemented in one state, Oklahoma. Oklahoma is, however, very representative of the United States. CNN Polling Director Keating Holland examined U.S. Census data and identified twelve key statistics—four that measure race and ethnicity, four that examine income and education, and four that describe the typical neighborhood in each state—and calculated how distant each was from the figures for the average state on each measure. Oklahoma ranked 6th nationally in the CNN poll for the most representative state in the country (Preston, 2006). Because of the representative nature of this state, it is probable that the K20 Model could be replicated in another state or nation that has a large number of rural schools that suffer from the inequities brought about by isolation.

Implications for Practice

The K20 Model for Systemic Educational Change and Sustainability advances social justice in rural schools through new conceptions of teacher professional development that enhances learning and prepares citizens for democratic participation. While equity in the United States is lacking for rural education, other countries may or may not provide greater resources and support to their rural schools. Additionally, in some countries, the geographical setting at which inequity occurs may be different—for example, in some countries suburban and urban settings may be short-changed (O'Hair and Reitzug, 2006). The K20 Model, however, is appropriate for addressing the disparities found in any setting. Both urban and suburban schools as well as the large number of rural schools in Oklahoma have been impacted by the K20 Model.

Addressing leadership first is one of the key implications of the K20 Model. A democratic leader is essential to the success of a democratic school. It is nearly impossible to conceive of a school functioning as a democratic learning community without the shared and supportive leadership from a democratic administrator (Hargreaves and Fink, 2006; Reitzug and O'Hair, 2002). In the experience of the K20 Center, many leaders want to enhance student learning, increase teacher efficacy, and expand family and community connections with their schools. However, they often lack the skills necessary to meet these goals. Through leadership training that models inquiry, discourse, equity, authenticity, leadership, and service, the K20 Center has addressed this deficit and empowered leaders to begin transforming their schools from traditional hierarchical leadership structures into democratic learning communities where all interested stakeholders take part in ensuring the success of each child within the school. Through the second phase of the K20 Model, leaders are offered the support of the K20 Center staff as they provide the needed professional development to help teachers, staff, parents, and, in some cases, students understand and begin to embrace the *IDEALS* of a democratic learning community. The use of technology as a motivational tool and an equalizer in this phase has seemingly accelerated the transformation of network schools (Atkinson, 2005; Williams, 2006; Williams et al., 2007). Teachers begin to collaborate and practice inquiry and discourse as they learn to integrate new technologies into their curriculum. Technology seemingly breaks down the barriers of departments, grade levels, or specializations, allowing teachers to work as a team whose concern is not curricular but rather the achievement of every child within their school. A second implication, therefore, of the K20 Model is that some sort of catalyst is required to allow teachers to meet as equals and concentrate on teaching and learning as a generic concept rather than focusing on their own specialization.

During the second phase of the K20 model, two further implications emerge. One important component of this phase is that leadership, or learning, teams are formed in each school. Staff from the K20 Center works with these teams and provides them with skills and structures necessary to allow them to build the capacity to function as leaders of their school in implementing changes that lead to democratic educational conditions. These teams are essential to ensuring the continued growth of the school after the K20 Center is no longer working with staff on a monthly basis. Such capacity building would be an important consideration for any program hoping to transform schools from traditional to democratic entities. In addition to teams of K20 staff members who work in each school, a single field technology specialist works with faculty and the learning team each month. This sustained training, designed to meet the needs of each particular school, builds the trust necessary to allow for meaningful discussions involving teaching, learning, and beliefs to take place. The trust engendered by this model is a further implication that should be considered in any school reform program. This trust is the essential foundation for the growth necessary to transform a school.

In the third phase of the K20 model, individual teachers begin to look at their curriculum and their pedagogical techniques. The trust developed in the previous phase is key to the success of this program. Because teachers are asked to change the way they teach, trust is an essential component. Again, the implication of this model is that sustained relationships between schools and reform agencies must be established if meaningful change is to occur.

During the final phase, the K20 Model works to enhance student learning by utilizing cutting-edge technologies. Through digital game-based learning, students increase curricular knowledge while engaging in an activity that seems to be play. The challenge of this phase for the K20 Center was ensuring that curricular content was sufficient to impact student achievement while at the same time creating a learning environment that students would find engaging and challenging. A confluence of knowledge and skills was utilized to meet this challenge. Educators served as subject matter experts to create the content of the game while undergraduate computer programming students built the gaming platform. The implications for other programs, which could be drawn from this phase, are that divergent models must be considered when attempting to reach today's digital learners. Schools must be willing to explore other avenues for delivery of content knowledge since the students they are educating today differ considerably from the ones they educated ten years ago.

The overall implication of the K20 Center Model for Systemic Educational Change and Sustainability is that a well-articulated, longitudinal approach is required if schools are truly to be transformed into democratic learning communities. A one-shot professional development program will not achieve this goal. Sustained and ongoing activities based on relationships built on trust and utilizing the most efficient tools at our disposal must be undertaken to transform schools. Such a complex and in-depth program may seem daunting, but the need for all students, in all settings, to be provided with the best possible education demands that we make such a commitment to democratic learning in our schools.

Further Sources of Information

Cate, J. M., Vaughn, C. A., and O'Hair, M. J. (2006) "A Seventeen-Year Case Study of an Elementary School's Journey from Traditional to Learning Community to Democratic School." *Journal of School Leadership*, 16(1), 86–111.

O'Hair, M. J., McLaughlin, H. J., and Reitzug, U. C. (2000) *Foundations of Democratic Education*. Fort Worth, TX: Harcourt Brace.

O'Hair, M. J., Reitzug, U. C., Cate, J., Averso, R., Atkinson, L., Gentry, D., Garn, G., and Jean-Marie, G. (2005) "Networking for Professional Learning Communities: School-University-Community Partnerships to Enhance Student Achievement." In *Network Learning for Educational Change*. Ed. W. Veugelers and M. J. O'Hair. London: Open University Press.

Reitzug, U. C. and O'Hair, M. J. (2002) "From Conventional School to Democratic School Community: The Dilemmas of Teaching and Leadership." In *School as Community: From Promise to Practice*. Ed. G. Furman-Brown. New York: SUNY Press.

Veugelers, W. and O'Hair, M. J. (eds.) (2005) *Network Learning for Educational Change*. London: Open University Press.

COMPONENTS	PURPOSE	KEY CONSTRUCTS
Authentic Research Experiences (ARE) engage teachers in scientific discovery with university scientists and assist them in teaching for conceptual understanding.	ARE provide teachers sustained opportunities to experience an instructional strategy while they study, experiment with, and receive helpful advice on content; collaborate with professional peers both within and outside of their schools; have access to external experts (that is, scientists); and have influence on both the substance and the process of their professional development (Slater and Cate, 2006; Newmann, 2000).	ARE include (1) an introduction to scientific research, research methods, and experimental materials; (2) the designing and conducting of original authentic research with guidance from scientists; and (3) the translation of research experiences into classroom practices that focus on conceptual understanding and problem-solving skills.
Lesson Study, originating in Japan, is an iterative process focusing on what teachers want students to learn rather than on what teachers plan to teach (Lewis, 2002; Yoshida, 1999). The lesson study process permits teachers to examine and adapt their practice, resulting in authentic achievement for students (Stewart and Brendufer, 2005).	The purpose of lesson study is not simply the improvement of a single lesson. Rather, lesson study provides teachers with an opportunity to examine their teaching in a way that results in the transfer of new knowledge acquired during ARE directly to their classrooms, ultimately resulting in the improvement of student achievement (Lewis, Perry, and Hurd, 2004).	A group of teachers develops a lesson together; one teacher teaches the lesson while others observe student learning; the group reconvenes to debrief, analyze, and if needed revise the lesson to incorporate the observations; and the teaching process begins again with a new teacher.
Digital Game-Based Learning (DGBL) enhances students' levels of engagement (Ravenscroft and McAlister, 2006) which has been associated with achievement (Inal and Cagitay, 2007). DGBL allows students to discover a new world, where they can be engaged in meaningful experiences that require a mixture of social interaction, critical thinking, theory testing, and evaluation (Inal and Cagitay, 2007; Lim, Nonis, and Hedberg, 2006; Mitchell and Savill-Smith, 2004).	DGBL creates opportunities for students to build their own models and test their hypotheses in the learning environment coupled with opportunities for reflection-in-action and reflection-on-action (for example, Schön, 1993). The models that students build are also used for assessment purposes to track development of students in their capacity to solve complex problems, which require higher-order cognitive skills (Eseryel, 2006).	DGBL experiences for teachers provide sustained opportunities for them to experience this instructional strategy (just as their students) while they study, experiment with, and receive helpful advice on interactive instruction resulting in modeling designed to develop higher-order cognitive skills. Teachers who have not experienced interactive instruction (for example DGBL) are ill-equipped to employ these instructional strategies in their classrooms (Newmann, 2000).

Figure 1.4 Continued

Professional Learning Community (PLC) is an approach to engaging school staffs in meaningful learning. PLCs create a collaborative, sustaining culture to improve the school's capacity to help all students learn at high levels (Dufour, Eaker, and Dufour, 2002; Huffman and Hipp, 2003; Lee and Smith, 1996).	Research reveals that a strong sense of community not only increases persistence, but also enhances information flow, learning support, group commitment, collaboration, and learning satisfaction (Rovai, 2002; Wellman, 1999).	Schools functioning as a PLC include the constructs: (1) creating a shared vision for teaching and learning; (2) promoting collective responsibility for student learning; (3) developing norms of collegiality among teachers associated with higher levels of student achievement (Averso, 2004; Lee and Smith, 1996; Little, 1993; Louis, Marks, and Kruse, 1996).
Service-Learning connects course content and virtual environments (that is, DGBL) with real world problem solving which requires students to identify, analyze, and help solve problems in their local communities. Service-learning has a positive impact on a student's sense of personal efficacy, identity, spiritual growth and moral development (Rockquemore and Schaffer, 2000).	Service-learning positively impacts students' (1) cultural and racial understanding (Vogelgesand and Astin, 2000); (2) sense of social responsibility and citizenship skills (Keen and Keen, 1998); (3) academic learning and the ability to apply learning to real-world problems (Foreman, 1996; Vogelgesang and Astin, 2000); and (4) career development (Astin and Sax, 1998; Tartter, 1996).	Service-learning positively impacts both students and the community by providing (1) real-world learning experiences linked to community needs (Budge, 2006; Furman and Gruenewald, 2004); (2) students with the skills necessary to exercise their rights as democratic citizens (Gruenewald, 2003); (3) a way for local communities to sustain their ecology and culture (Woodhouse and Knapp, 2000).

Figure 1.4 K20 Embedded Professional Development for Teachers

Chapter 2

Democratic Schools in Latin America? Lessons Learned from the Experiences in Nicaragua and Brazil

Silvina Gvirtz and Lucila Minvielle[1]

Participation and Democratization as Alternatives to Bureaucratic Governance in Education

From the end of the nineteenth century until the middle of the twentieth, national education systems, where the State played an essential role, consolidated. Within a "state-centric" logic, centralized and bureaucratic education systems were developed amid a technical expertise environment providing little participation to citizens in the decision-making process. This peculiar organizational design established autocracy and limited the features of citizen participation that later on became naturalized.

Nevertheless, during the second half of the twentieth century, the social, political, and economic context in which education systems developed abandoned its simple and static nature to become dynamic and complex (Mintzberg, 1993). Therefore, bureaucratic, centralized, and autocratic designs started to be questioned as to their capacity to offer quality education for everyone (Murphy, 2000). These issues led to consecutive reforms resulting in concrete policies in the 1980s and 1990s. These reforms took place both in developed countries, such as England or New Zealand, as well as in developing ones, such as South Africa or El Salvador.

In Latin America, the rediscovery of democratic forms on the fall of the military regimes governing the countries in the region served as the ideological foundation for these reforms. These transformation processes were based on the idea that the democratization of education systems would be accomplished by a wider use of voice[2] as a means of channeling the demands of the members of the educational community. At the same time, these policies supporting community participation in the schools' decision-making process aimed to promote learning about democratic practices among citizens, who bore the responsibility of developing young Latin American democracies. Local community members were at the centre of these learning practices, with special focus on the "educational community." It should be highlighted that these reforms defined the "educational community" as the ensemble of actors that included both expert members (employees of the state bureaucracy, principals, and teachers) and laymen (parents, students, and schools' neighbors). Here lies an important difference with the traditional education governance structures. Facilitating participation by nonprofessional members of the society in the schools' decision-making process (both administrative and pedagogic) constitutes an innovation that is at odds with the established school culture that privileges the pedagogic expertise when deciding who is qualified to participate in the decision-making process of education.

In Latin America, policies that intended to foster participation of members of the community in the schools' decision-making process took many forms and originated within different ideological postures. School choice by parents, involvement of parents and students in school evaluation commissions, participation of teachers, students, and parents in the election of the school principal, and parents, students, and teachers participating in the school councils were some of the concrete ways these policies found to allow nonprofessional members' voices to be channeled.

The participation of nonprofessional members (parents and students) in school governance bodies, which these reforms put forward, was intended to bring about fundamental changes in three directions. First, allowing parents and students to partake in school governance decisions was intended to modify the "who" that is entitled to make decisions within the school. Second, these reforms tried to alter the "what" of school governance, that is, the nature and the number of decisions that these lay actors could be involved in. And third, these alternative institutional arrangements looked forward to changing the "how" of school governance—that is, in the way in which governance decisions were carried out in the school.

Reference should be made to the conditions in which the above-mentioned reforms work in developing countries. In such countries, governance issues are frequently more complex and difficult to work out since problems of infant malnutrition, adult analphabetism, and poverty have not yet been solved. In the context of these problems, it is particularly interesting to study school governance in these developing countries. Two countries have been chosen for study, Nicaragua and Brazil, both representing, with their own peculiarities, useful examples of what is hereinafter described as a complex economic and social situation.

Nicaragua constitutes an interesting case since major governance and financial reforms were implemented in its education field in the past twenty years. The Autonomous School Program promoted by the right wing coalition governing the country at the beginning of the 1990s set forth radical changes at a national scale in the way schools were managed. Controversial measures such as parental cofinancing of some school activities and the widening of community participation were heavily criticized by opposition leaders from the left. Amid this political upheaval the program was implemented, and it stood in place for sixteen years until the left-wing party led by Daniel Ortega assumed power in late 2006 and abolished the Program completely in 2007, with the intention of implementing a new reform, which at the time of writing is yet to be formulated.

The Brazilian cases contribute to the objective of showing how these reform projects were conceived and implemented in very different political, ideological, and geographical settings. In the state of Minas Gerais, the Projeto Pro Qualidade, designed to be implemented in a whole state, was financed by the World Bank, which strongly supported and contributed to the project's implementation and evaluation phases. This project was part of a major reform initiative at the state level inscribed along the lines of rationalization and decentralization of state offices set forth by right wing administrations in the 1990s. On the other hand, the case of Porto Alegre was implemented in the municipal system of education (beneath the state level) and is an example of a reform implemented by a left-wing party in office. Both reforms were implemented in the first years of the 1990s and are still in place.

Making use of institutional theory as an analytical tool, we shall aim to analyze how far these reforms have gone in practice in fostering actual participation in school governance and generating democratic practices among the local community.

The Case of Nicaragua[3]

Nicaragua is a country of 5.2 million inhabitants with a GDP of US$4,911 million, equivalent to an annual amount of US$950 per capita (World Bank, 2006). By 2007 end, 23.3 percent of the population aged over fifteen years was illiterate, more than 20 percent of the population did not have access to improved water sources, and 10 percent of children below five years was underweight. As a consequence of these and other indicators, the country ranked 110th in the Human Development Index formulated by the United Nations Development Program (UNDP, 2007).

In 1991, after many years of a socialist government, a center-right coalition assumed the executive power after defeating the Frente Sandinista de Liberación Nacional (FSLN) in the presidential elections. This government, whose leader was Violeta Chamorro, implemented economic deregulation policies and a general state downsizing plan. In the area of education, there was a significant reduction of the educational bureaucracy, through which the Education Ministry staff was reduced by 50 percent in less than two years. Decentralization strategies were implemented for municipalities as well as for schools. Within this framework, the policy outline called the Autonomous Schools Program (ASP) was developed and implemented.

The ASP basically consisted of increasing the local communities' rights vis-à-vis central bodies to decide on schools' academic and pedagogical matters. To carry out this policy, a school board or managing council, called Consejo Directivo Escolar (CDE), was created as the school's most important governing body. The purpose of this institution, according to the explicit reform objectives, was to establish a body in charge of channeling community members' educational demands, thus allowing them to participate in a school's life. Therefore, according to the norms established by the Program, representatives of all the members of the school community constituted the CDE, which included the school principal and representatives of teachers, parents, and students. At the time our field research was carried out, the ASP had been implemented in 1,780 schools, representing 63 percent of total students in the system and 55 percent of the teaching force. As of 2006, the implementation of this school governance structure was required in all public schools. In 2007, when the administration of the left-wing president Daniel Ortega took office, the ASP was to be dismantled completely on account of serious accusations of malfunctioning. A new reform is yet to be designed and implemented. In the light of these developments, the analysis of the reform itself becomes more necessary and interesting than ever before.

One of the main features distinguishing the ASP from the traditional bureaucratic-centralized government design of Nicaraguan schools was its decentralization. According to the Program's aims, schools belonging to it had broader powers to make academic and administrative decisions related to both daily operation and long-term issues. For example, the school had the right to decide on the length of the school day and schedules for the school year, review and approve internal standards, and verify their fulfillment with respect to national standards.

Decentralization of the decision-making process seems to be a good starting point for a reform program seeking to move the decision-making process nearer to members of the local community and to foster the learning of democratic practices. However, decentralization is a necessary but not sufficient condition to ensure local members' participation. This being the case, in order to verify to what extent the ASP permitted every member of the local community to channel his/her voice, we centered our analysis on a micro perspective that allowed us to focus on what happened inside schools. Our study was then centered on the school governing body in order to verify in everyday practice "who" made "what" decisions, and "how" these decisions were carried out.

A first analysis of the laws, rules, and regulations regarding the Program's operation shows that the CDE, being the school's local governing body, had more power to make decisions in the administrative realm than in the academic. This was so because a considerable number of academic decisions, such as the definition of curricular contents, were maintained, even after the reform, within the realm of the National Ministry of Education.

Studying formal regulations is a necessary step in any reform program analysis. However, the micro perspective adopted in our research shows that examining everyday practice in the schools we visited is more revealing about the effective participation of members from the local community and the possibility of generating learning for democratic practices than the formal analysis of regulations.

One of the first facts we encountered when studying actual practice in schools was that, although the law promoted the participation of all the members in the educational community, in practice there was little participation from nonprofessional actors (parents and students) in the CDE. As a general rule, few candidates applied, and once elected, they rarely attended the council's meetings.

Two types of explanations for this phenomenon were drawn from our interviews with principals, teachers, parents, and students. First, the meager participation from nonprofessional members of the educational community in the CDEs was said to be related to their allegedly low intellectual level. According to our interviewees, due to their low intellectual level and lack of practice in democratic decision-making, both parents and students found it difficult to understand how important their participation in school governance was, either in the immediate impact on the school's decision-making, or in the long-term development of citizen skills arising out of the implementation of democratic practices.[4]

Second, lack of attendance of nonprofessional members at CDEs was said to be related to the difficulty of implementing participative governance structures in social and economic contexts typical of developing countries. Attending CDEs' meetings demanded time and dedication. These are requirements that parents of low-income families who must struggle every day for (mere) economic survival found it difficult to meet.

In a personal interview with Lic Zepeda, director of ANDEN, the most important union in the country, he firmly stated this point:

> The population of Nicaragua has great difficulties. The main one is that most of them are trying to survive day after day. So, most of them are thinking how to live day by day and not thinking about their rights. Therefore, survival deprives them of the right inherent in being a citizen. (Personal interview—Managua, March 2004)

In order to operate properly, this reformed school governance arrangements required all members of the CDE (professional and laymen) to be regularly present and involved. However, this did not occur among those entitled to making decisions, especially parents and students. Therefore, the democratic reform of Nicaraguan schools faced great difficulties in generating democratic learning since frequently the ensemble of actors taking decisions within the school did not seem to represent the interests of all the stakeholders in the educational community.

For the purpose of completing the analysis on the effective existence of democratic procedures within the school governance and the possibility of generating an active learning for democratic practices, it is necessary to study in detail how the decision-making process operated within the CDE, that is, "how" these decisions were implemented in this governing body.

The field study revealed that even though formal rules granted the same amount of power in making decisions to both professional and nonprofessional members, representatives of parents and students wielded little influence in the resultant final decisions taken within the CDE. In other words, in practice there was an unequal distribution of power among council members, clearly favoring professional ones. The cases studied showed that it was generally principals and teachers who suggested and decided on the matters being discussed, and, at the most, decisions were ratified by nonprofessional members.

This modus operandi set a participation style for nonprofessional members of the CDE that we describe as "peripheral" or "passive." Most of the participation by parents and students in schools that were part of the Program generally took place outside the CDE in volunteer tasks, such as assisting in the school dining hall or organizing fairs, always under the close inspection of professional members. The peripheral role played by nonprofessional actors in the CDE did little to help generate practices resulting in a learning for democracy, since, as Dahl puts it, democratic practices require that everybody has equal and appropriate opportunities to express their preferences and to put forward reasons in favor of or against proposed positions, thus having the opportunity to produce an effect on the final decision (see Dahl, 1994).

Finally, in relation to the role played by nonprofessional members of the CDE, the field study revealed that subordination of parents and students to teachers and principals had a double origin. On one hand, professional members encouraged their own situation of superiority. The following extract from an interview with a teacher clearly explains this viewpoint, which was present in almost all the interviews we carried out with teachers and principals:

> Sometimes, parents try to give orders more than the headmaster and other similar situations are raised. The headmaster is the Ministry of Education's representative to put forward ideas on pedagogic procedures, so a parent cannot go over the headmaster's authority. (Teacher, Autonomous School 14 de Septiembre- Personal Interview, Managua, March 2004)

On the other hand, the peripheral type of participation found in Nicaraguan schools was also justified by the perceptions of parents and students about their own abilities and of their rights to belong to the CDE and make governance decisions. When queried about the reasons for this kind of peripheral participation, nonprofessional members stated they were not "knowing" enough about schooling in order to participate in the decision-making process. For them, the natural thing to do was to remain in a passive role, leaving active decision-making to the schools' professionals. Such type of reasoning based on the Competence Criterion (see Dahl, 1994), which states that only those with technical knowledge should be entitled to partake in any decision-making process, seemed to be a part of the deeply embedded basic assumptions that constitute school culture. Such cultures are resistant to the implementation of institutional designs that foster learning for democracy.

We discussed so far about "what" was decided in the CDE, "who" were responsible for making the decisions, and "how" this decision-making process operated. The results of the analysis of the regulations were widely different from the results of the analysis of everyday practice. From a formal point of view of the regulations, an alternative governance practice was to be generated, opening participation to nonprofessional members in school decisions. However, learning for democratic practices, that is, learning intended to make citizens familiar with democratic discussion and decision-making, has been threatened by the continuous use of traditional practices deeply settled in school culture, whose main actors were the holders of what was seen as expert knowledge, leaving little space for professional members and laymen to interact genuinely.

The Case of Minas Gerais

Minas Gerais is Brazil's fourth largest state in terms of inhabitants, with a population of 19.2 million, of which 82 percent live in urban areas. It is considered a transitional state between the poor northeast and the rich southeast. Minas Gerais has 853 municipalities that show a great inequality in income distribution

The basic education system (which includes basic and secondary levels) has 4.7 million students, of which 90 percent attend public schools. Among these schools there are practically no federal schools (i.e., schools that belong to the national school system): most students attend state or municipal schools (52 percent and 38 percent, respectively) (Instituto Brasileiro de Geografia e Estatística, 2006).

With respect to reforms carried out across the country, in the 1990s the state of Minas Gerais put into practice a general state reform program based on a model that pursued simplification, debureaucratization, and decentralization. During this decade, Minas Gerais's education policy involved a series of different general measures that were implemented simultaneously, with financial and technical support from the World Bank through a global program called "Pro Qualidade." With the explicit purpose of democratizing education and achieving quality education levels for all, one of the program's main lines of action was including members of the community in the schools' decision-making process. In order to achieve this goal, two measures were implemented: the democratic election of school principals and the reincorporation of school boards, called Colegiados Escolares (CEs), which had been created in the early 1980s and had been gradually disappearing ever since.

The CEs were quickly put into practice, perhaps because of Minas Gerais's previous experience. In 1997, only five years after the project's implementation, school boards were operating in all public schools (of primary education level) within the state education network.

One of the first curious aspects of Minas Gerais's reform is that, although it explicitly aims at democratizing the system by granting more power to the members of the local communities in taking educational decisions, the system's regulations allow only a few decisions to be carried out locally. According to the rules mentioned above, the school, as a local decision unit, is empowered to act in matters of a lesser degree of importance, concerning only details and not any substantial matters. As an example, although the school formally participates in preparing the curriculum, there are very few decisions about the curriculum that it is empowered to make. For example, while schools are granted the power to order the teaching contents of the curricula for the school year (i.e., they can decide *when* to teach the different subjects of the curricula defined for that school year), members of the central bureaucracy decide on the content itself (i.e., they decide *what* is going to be taught in each school year). Bureaucrats get to decide even on the pedagogic methods to be used to teach those contents in the classrooms. This limited scope of action that the school has as a local decision maker interferes with the possibility of generating opportunities so as to foster the learning of democratic practices in the school and its community.

The possibility of generating a space for genuine local participation within an alternative framework of a democratic decision-making process appears even more confined when analyzing those decisions that are indeed allowed to be taken locally. In this context, the school, even when formally empowered to make decisions, is considerably limited by the supplementary regulations passed by the central bureaucracy. These supplementary regulations, about the way schools should proceed, are of such a precise and exhaustive nature that the decision-making process in schools looks more like the application of standardized processes than a real selection among alternatives where different local stakeholders have the possibility of expressing their preferences and discussing the options until a final decision is reached.

The management of school funds serves as a good example of this *fictional* power to make decisions. According to the general law of the Brazilian educational system, the *Lei de Directrices e Bases (LDB)* (Law of Guidelines and Foundations), and Minas Gerais education laws, schools are empowered to manage their funds through two governance bodies, the CE and the "Caixa Escolar" (School Cash Box), the latter being an ad hoc body created to deal with fund management. In practice, supplementary rules and decrees determine that schools and these governing bodies have very little space to make decisions on these matters. In Minas Gerais, schools have three sources of funds. The first one comes from the federal government. These funds are earmarked for specific purposes. Thus, schools' governance bodies have no power to decide upon them. The second source, which is the most important in terms of resources, derives from the state education department. The school is entitled, by state regulations, to manage only 8 percent of these funds, which are generally used for day-to-day minor expenses. The remaining 82 percent is allocated by the Secretaria Gradual de Educación de Minas Gerais. Finally, schools in Minas Gerais have self-generated resources, coming from activities such as the renting of school premises, school kiosks, and the sale of raffles. Over this last source of funds, schools do have the power to decide on its allocation, but these auto-generated funds are generally very small and contingent on schools' generating the resources. This funding scheme seriously limits the scope of decision-making that schools have, showing an important discrepancy between what the law mandates and what really happens in practice.

Further reform in Minas Gerais, aimed at democratizing schools, is likely to face a very challenging task, judging by the results of our analysis of the participation of both professional and nonprofessional members in the CE. The practice of participation has not created significant possibilities of generating a genuinely alternative governance structure. The CEs of the schools that were visited show little presence of nonprofessional members. According to the interviews we conducted with professional and lay members of the school community and the schools' records we analyzed, parental attendance, either by their representatives or in a direct manner, as for example in General Assemblies, where all parents are expected to come, is very low. The following statement of a principal from one of the program's schools clearly puts forward this point:

> When the time of elections arrives, we announce and call the community and a relative attendance is achieved. Not in the degree we expected for in a population of 5,000 parents, sometimes appear 70 or 100. It is something, for in the past, nobody appeared. (De Oliveira, 2002).

What leads to this low presence of nonprofessional members in the CEs of Minas Gerais? The interviews carried out revealed that the main reason behind the apparent apathy of parents of the community is that they consider participating in matters related to the decision-making process in schools as a high-cost activity. That is, they believe that it implies a great amount of time and effort. At the same time, the perceived benefits in return are few. In other words, participating in the meetings of the CE and spending a considerable time in discussing matters related to details instead of key matters is not attractive for the majority of those interviewed, that is, mothers and fathers holding one or several jobs, or having a family to be looked after.

The above-mentioned scenario resembles that of the Nicaraguan case with regard to how decisions are made within the CE. The small space left for taking decisions, related only to matters of lesser importance (such as deciding upon the use of funds for minor day-to-day expenses vis-à-vis managing the school's budget), together with the importance given to the expert knowledge of teachers in the Latin American school culture in general, and in Minas Gerais in particular, leaves the path open for decisions to be made autocratically by professional members of the school, that is, principals and teachers.

In summary, the incorporation of democratic practices in schools and their communities, considered as one of the main aims of the reform program, has had little or no presence in the schools visited. As in the case of Nicaragua, the persistence of practices seems to preclude an alternative form of school governance from developing in Minas Gerais, which struggles under the burden of school culture and, in this case, is permanently constrained by a group of regulations that seems to be conceived so as to limit any modification to the traditional form of governance.

The Case of Porto Alegre

Porto Alegre is one of the 496 municipalities of the Brazilian state of Río Grande do Sul.[5] With a population of 1.4 million distributed in eighty-five neighborhoods and an urban rate that surpasses 97 percent, it is the most populated municipality in the state. On the economic front, figures show that, although Porto Alegre is the municipality with the highest GDP of the state (around US$5,400 million), the GDP per capita is around US$3,800 (Fundaçao de Economia e Estatística, 2005).

The education network administered by the municipality (Rede Municipal de Educaçao (RME)) is small because most of the students attend state schools. The 69,000 students within the education municipal network attend ninety-two municipal schools generally located in marginal areas that are home to low-income families. The majority of the students attend the basic education schools, which are divided in three cycles. The first cycle serves children between six and nine years old; the second, students from nine to twelve years old; and the third, students from twelve to fourteen years old (Secretaria Municipal de Educaçao de Porto Alegre (SMED), 2007).

In 1989 when the left-wing administration, called the "Popular Administration," assumed power in Porto Alegre, during its first year of office it implemented several public policies geared toward democratizing the relationship between the educational community and the school. This resulted in the creation of the "Escola Cidada" (Citizen School). This program, which applied only to municipal schools, had three main objectives:

- the decentralization of financial resources from the municipality to schools,
- the democratic election of the schools' principals, and
- the creation of school boards, called Conselhos Escolares (CEs), which comprised representatives of the education community members (the school principal, teachers, parents, and school students) and were constituted as school governance bodies.

At the time our fieldwork was conducted (end of 2003) all the schools of the RME had their school boards working, had chosen their principals, and had received funds from the Municipal Education Department that they were entitled to administrate as part of the decentralization of financial resources initiative that was part of the Escola Cidada program.

A first look at the regulations for this governance structure suggests a system that is far from decentralized—that is, if the *number* of matters to be decided upon by the school is considered. For example, schools, unlike the previous cases of Nicaragua and Minas Gerais, are not allowed to control the evaluation process of their students, nor can they hire or fire teachers, or decide upon the schools' organizational structure. However, in order to understand the way in which the institutional design offered by this alternative contributes to the generation of democratic practices through local discussions, it is necessary, as in the case of Minas Gerais, not only to consider the amount of decisions the school can decide upon but also the *importance* of the decisions that members of the local community within the school are allowed to take.

This analysis shows that those few matters discussed in the school framework, both in their academic and administrative aspects, are very important for the school's life. For instance, schools belonging to the RME are empowered to discuss and decide, in consultation with members of the community, on the content, organization, and implementation of the curriculum.[6] They are also entitled to both appoint and dismiss the school principal, and to actively participate in the evaluation process of teachers.

This situation, where discussion is focused on a few, key matters, sets a clear difference with the case of Minas Gerais previously studied (where the few decisions under the control of the school community had little influence in the school's life). It stresses a key point already mentioned in the previous cases—it becomes necessary, upon studying how government structures work, to apply a micro perspective, which not only takes into account quantitative data but also considers the phenomenon from a qualitative viewpoint. It is only from this latter perspective that it is possible to assess if the alternative governance designs trying to generate learning for democratic practices achieve this in practice.

A second step toward the discovery of how the alternative governance structure in Porto Alegre actually works requires verifying who makes what decisions within the school. In order to bring about the possibility of fostering genuine learning for citizens within a framework of discussion and voting, it is necessary to achieve a broad and active participation of all members of the local community, both professionals and lay actors. The CE regulations state that all the CE's members have the same powers to decide on the matters assigned to the control of the CE. These matters pertain both to the administrative as well as the academic field. In contrast to the other cases reviewed, in Porto Alegre the institutional design explicitly intends to give nonprofessional members of the community (parents and students) the opportunity to express their opinions and take part in decision-making concerning matters traditionally resolved by principals and teachers.

In spite of this attempt to include all community members in schools' governance discussions, everyday practice at schools showed a different situation regarding lay participation. Due to the high importance granted to technical knowledge in school culture, in practice it is the perceived competence in the academic field that is the criterion by which all members of the educational community determine who actively participates in the relevant decisions at school. This leads to the exclusion of nonprofessional members from certain decisions (especially in the academic field). This particular situation is not only encouraged by teachers and principals but also by those same lay members excluded from the process (parents, in particular). The statement of one of the parents interviewed, acting as president of one of the CEs visited, clearly explains this situation:

> I am a lay person on pedagogic matters . . . all professors were formed in the university, . . . [in] the pedagogic field, . . . [under] supervision, so they know much more than me. As far as I am concerned, I give my opinion; but when I consider I am saying nonsense, I don't speak . . . [I]n the council, the person that doesn't know how to express herself, won't do it. (Personal Interview—Wenceslau Fontoura Municipal School. Managua, March 2004)

This viewpoint certainly affects the possibility of achieving effective participation by the nonprofessional members of the CE. In the councils of the schools visited in Porto Alegre, even though the physical presence of parent- and student representatives was higher than in the other cases studied, it was still noticeable how teachers and principals outnumbered parents and students and generally controlled discussions and decisions. This peripheral participation by lay members militates against achieving the context of discussion and interchange of opinions required by a democratic and inclusive decision-making process fostering learning for citizens' life in the local community.

Given the peripheral participation of lay members, our fieldwork revealed a curious phenomenon. Upon interviewing parents on whether they considered school governance to be democratic, the majority answered affirmatively, highlighting the fact that they indeed felt they belonged to an active school community. An apparent contradiction then arose between a (self) exclusion of lay members from important areas of decision in the school governance process, and the perception these same lay members had of the democratic nature of the decision-making process. What is then the explanation for this apparent inconsistency where, on one hand, competence criterion rules and lay actors intervene peripherally in the decision-making process, and, on the other hand, there is a generalized perception of an inclusive and democratic process in the municipal schools of Porto Alegre?

The key factor in order to understand this paradox is the existence of other participation opportunities, different from the CE, which are less restrictive when trying to gather parents and students together with professional school members, and whose approach is friendlier and more welcoming for lay members. Ad hoc commissions composed of parents and teachers created to determine admission criteria, or informal meetings between teachers and parents where tea is shared and strategies to cope with specific student problems are discussed, are much more comfortable examples of participation, inviting parents to explain their viewpoints and discuss until a decision satisfying the majority is reached. In these less structured and more informal collective participation spaces, the idea that the voice of the nonexperts shall not be taken into account does not rule. Parents are comfortable to express their point of view and, from the interviews we conducted, they seemed to feel in the right to do so, since they justified their participation in a different kind of knowledge, the kind that arose from the experience of being parents and having their children at school. In this way, a truly democratic decision-making process is achieved, based on the equal opportunity to participate for every member of the community.

These genuine, informal participation spaces in Porto Alegre are grounds for some optimism about democratizing schools there. Even though the basic assumptions of school culture do not contribute to the generation of learning for democratic practices, they show that through trial and error, there are ways in which the true political will of democratizing the system may open the long path of finding the appropriate institutional designs to achieve the objective of genuinely generating an alternative for participative governance.

On the (Real) Possibilities of Generating Learning for Democratic Practices

During the last quarter of the twentieth century, Latin America has gone through a process of gradual democratization. This led to the need to develop, in the young democracies of the region, a government culture based on citizens' participation, discussion, and joint decision-making.

Reform policies in education governance structures described in this chapter have been framed within this context, pursuing, among other objectives, the generation of learning for democratic practices by means of opening to members of local communities opportunities to participate in the schools' decision-making process.

The cases studied reveal a big gap between the alternative offered by these reforms and what is effectively achieved every day at schools. Even when, as in all three cases, the regulations tried to generate learning for democratic practices among community citizens, by involving nonprofessional members in the discussion of school decisions, the field study showed that parents and students' participation in these activities was low and peripheral; their commitment was weak and their subordination to professional members of the councils (both teachers and principals) was high.

In consequence, a substantial lack of discussion in the decisions of the school councils was noticed. The gap left by the absence of discussion was generally occupied by the technical knowledge of teachers and principals, especially in the academic field.

Our conclusion is that the reform projects we have studied have a long path to follow in order to achieve the democratization of education systems. However, the short distance already traveled reveals certain insights generated by these experiences. It is with a focus on these that new alternatives for the future should be designed.

What are the lessons learned? Four lessons are suggested that concern the design of reforms that play an important role in effectively achieving a feasible alternative for democratic governance at schools.

Enhancing the Benefits and Diminishing the Costs of Participation

The first point is that the institutional design of the reform plays an important role, particularly the benefits and costs that it generates for participants.

One of the suggestions to increase both the physical presence of members of the local community and the intensity of their participation is to enhance the benefits as perceived by nonprofessional participants. One way to do this, which at the same time contributes to the main objective of democratizing the decision-making process in schools, is to grant to the school governance bodies, through their representatives, the power to make decisions that have a significant impact on the lives of both schools and students. It is more likely then that parents and students will be motivated to dedicate to these activities their time and effort, which are generally scarce, especially in cases where parents and students have to deal with the additional problems and sacrifices of poverty.

Another way of working in the interests of achieving an effective participation of all the members of the local community is to diminish the participation costs for the nonprofessional members of the councils. Considering that they have several obligations, apart from that of participating in the decision-making process at school, it is necessary to regulate the intensity of demands in terms of time and frequency.

The Importance of More Informal Opportunities for Participation

The second point, apparent in the case of Porto Alegre, concerns the benefits of offering other participation opportunities beyond involvement in the school councils, such as informal meetings or short-term committees for specific purposes. These are especially useful in local communities with little experience and with a need to learn or develop new discussion practices and decision-making processes. These spaces, opened to the community, allow for less structured environments for those who lack experience in education matters, thus making them feel more at ease and helping them to question and abandon the deeply settled assumption that only experts can participate in education.

The Need for Deep Change in School Culture

The remarks in the second point lead to the third point, which may be the most important discovery of the present study on alternative school governance designs and the possibility of generating new learning for a democratic life. The cases studied showed that institutional culture constitutes a key element for either the success or failure of the implementation of alternative ways of doing things within the schools. In the three cases studied, the basic assumptions deeply settled in the school culture, which reflect the importance given to the technical expertise of both teachers and principals, permanently hamper lay actors' participation in the decision-making process. As a result, there is a shortage of discussion and a submission to dominant conceptions of expertise that inhibit the application of democratic practices within the school.

Following this line of thought, it is suggested that the degree of difficulty to be faced by school governance reforms will largely depend on the depth of the change taking place in the institutional culture. It should be noted that if a deep change in school culture is not achieved together with the governance reform, then participation by nonprofessionals will be very difficult to accomplish. This being so, fundamental change in institutional culture is needed in order to have satisfactory results when implementing these kinds of reform.

The deeper the change being sought, the more we need to analyze and discuss the basic assumptions ruling the school culture. In order to achieve real possibilities for generating learning for democratic practices through the participation of members of the local community in the decision-making process in school, it is indispensable, before implementing any reform project of this magnitude, to study these underlying assumptions and to discuss how they operate in the school logic. It is only in this way that institutional designs will be created, which acknowledge that learning to be democratic will not be rapidly achieved since it requires a consensus that can only be developed in the long term.

The Need for A Micro Perspective in Designing, Implementing, and Revising Reform

This previous observation is especially meaningful in the light of the past events taking place in Nicaragua, where the reform was abolished altogether, and leads to our fourth point. Stressing an observation that appeared repeatedly in our work, we would like to conclude by stating that school governance reform projects, even when working on a national scale, should adopt a micro perspective when being designed, implemented, and revised. Reforms can be conceived, planned, and designed as macro policies that use the education system as a unit of analysis. Nevertheless, implementation occurs (or does not occur) at the school level. School community members are the ones who every day decide (or not) to put into practice policies designed by the education bureaucracy. This is why we claim that it is only by looking into what really happens inside the schools that reforms will be able to have a chance to be institutionalized and become part of the school culture.

Notes

1. This research is the result of an investigation carried out in 2003 and 2004 with a subsidy granted by the John Simon Guggenheim Memorial Foundation.
2. We use the term "Voice" as opposed to "Exit," which is another way of channeling demands in Hirschman's theoretical Framework (Hirschman, A., 1970).
3. Our study consists of analysis of documentary sources and personal interviews. Documents that were used as sources of data include laws, decrees, and regulations concerning the educational field. On the other hand, a total of forty interviews were carried out. Interviews took place in educational ministries, State education departments, and municipal secretariats, and eight of the schools. Actors interviewed include school principals, teachers, parents, students, and members of the educational bureaucracy and academic establishment, such as ministry officials, teachers' union representatives, and university professors.
4. When interviewing Dr. Juan Bautista Arrien, president of the UNESCO office in Nicaragua, we got a good example of this line of thought:
   "Interviewer: Do parents participate?"
   Arrien: "There it is, that is one of the problems. Fathers and mothers are not qualified in many of the aspects related to education, pedagogical interaction . . . we

have to recognize that parents do not have the intellectual development [to participate in school governance]." (Personal Interview, Managua, March 2004)
5. Rio Grande do Sul is known as the "gaucho" state, and although it is a relatively small state in size (only 3 percent of the country of Brazil), it is one of the most important states from an economic point of view. With 10.5 million inhabitants and a GDP of more than U.S. $31,000 million, it is the greatest grain producer in the country and the second commercial and industrial pole of Brazil. One of its achievements is the best Human Development Index in the country.
6. The process of curriculum definition by schools is described in detail in Gandin and Apple (2002, 267). In the authors' words: "The starting point of this new process of knowledge construction is the idea of thematic complexes." Through action research (one that teachers do in the communities where they work, involving students, parents, and the whole community), the main themes from the specific community are listed. The most significant ones are constructed in the thematic complex that will guide the action of the classroom in an interdisciplinary form during a period of time. The traditional rigid disciplinary structure is broken and general interdisciplinary areas are created.

CHAPTER 3

THE TOUCHING EXAMPLE OF
SUMMERHILL SCHOOL

Ian Stronach and Heather Piper

Introduction

Summerhill School was founded in 1921 by A. S. Neill, whose work on education and child development was of international repute, especially in the 1960s and 1970s when it became a "transatlantic cult" (Skidelsky, 1969, 15); his publications have never been out of print in Japan and are now being reprinted in the United States and UK (Ayers, 2003; Vaughan, 2006). Neill was interested in practice not theory, and Summerhill was his exemplar. The school currently describes itself as "the oldest child democracy in the world" (www.Summerhillschool.org) and remains unchanged in its ways of self-government since Neill's time. It is a predominantly residential "free" school, one much inspected and criticized down the years by the relevant Government Inspectorates in England. In 1999 the Education Inspectorate (Her Majesty's Inspectorate (HMI) who work within the Office for Standards in Education (OfSTED) tried to close down the school, lodging a series of objections that the school was forced to legally resist in order to remain open and true to its principles. Ending the policy of voluntary attendance at lessons has been the government's enduring target. The government failed, and since then the school has prospered. Indeed, a recent Social Services inspection report praised the school's "very high" levels of pupil satisfaction (CSCI, 2005). The conservative *Telegraph* newspaper contrasted today's Summerhill with CSCI criticisms of leading English private schools: "some boarding schools may be wondering today whether A. S. Neill was not on to something after all" (Claire, 2005).

The current roll (around 90 pupils) includes children from age four to sixteen, from countries as varied as the UK, United States, Germany,

Holland, Japan, Taiwan, and Korea. The core of the school is the Meeting, where pupils and staff, on a one-person-one-vote basis, decide how the school will be run. In addition, younger children may also appeal to older children (this varies according to maturity, but is generally for those over eleven) who act as mentors and are called Ombudsmen (of either sex). A series of committees and pupil-appointed functionaries run aspects of the school—for example, bedtimes are decided by the Meeting, and are supervised by "Beddies Officers." The school is divided into age-related houses, called Cottage, San, House, Shack, and Carriage, in ascending order of age. Spatially, the grounds comprise several acres of woodland, some open grassy areas, a large house, and a number of single-storey classroom blocks. Most staff live in and are housed in caravans scattered around the grounds. The school is located on the edge of a small town in rural East Anglia, England, where it was relocated in 1927. The school is fee-paying and has charitable status. It has affinities with Deweyan schools, in its child-centered rationale and emphasis of democratic government, and served as a model for over 600 such free schools in the United States in the 1960s and 1970s. To the school's knowledge, none of these offshoots now exists, although of course a few related "free schools" survive.

Research Evidence

The easiest way to interpret Summerhill is to succumb to its difference. It is democratic, while schools are generally autocratic. There is an egalitarian relation between adults and children. It rejects compulsion in relation to attendance at lessons, examinations, assessment, and even reports to parents. It is exotic and so we read it, easily or uneasily, against our prejudices.

The research we refer to throughout is a single element of a larger project focused on the controversial issue of "touching" behaviors between education and childcare professionals and children in a number of settings.[1] We had become aware that the "touching" of children, as an aspect of professional practice, was causing concern: "Physical contact could be misconstrued by a child, parent/carer, or observer . . . it is unwise to attribute touching to the style of work or way of relating to pupils" (primary school teacher, Piper, 2002, 2). Reported injunctions, with which we became familiar, included having a second adult to witness intimate care routines, minimizing cuddling young children, and even requiring particular ways of doing this, such as the sideways cuddle (to avoid full-frontal contact).

Summerhill School was selected because we anticipated,[2] on the basis of previous research experience in the school, that it could be seen as being at one end of a continuum, a school generally thought to be less "regulated" and where pupils were part of a self-governing community. The original intent to explore "touch" at Summerhill was quickly affected by the recognition of touch per se as a banal issue. The unique characteristics of Summerhill made it obvious that "touch" is merely an aspect of other social, organizational, cultural, and ethical features of the environment in which it occurs. We could only understand "touch" at Summerhill if we explored Summerhillian practices of the self, the other, the community, the culture, and their conjoint reproduction.

As we asked about adults touching children, children touching each other, adults touching adults, it felt a bit "pervy" as a subject for conversation, an attempt to unnaturalize what the subjects regarded as absolutely normal. Our discomfort surfaced in our field-notes: how do you say they "rubbed against one another" (two Taiwanese girls) or "he put his hand on her thigh . . . without immediately being in a sexual register?" What was the overall source of our uneasiness? It seemed that asking such questions carried a sexual innuendo that became more prominent as the fieldwork continued. In a 2001 mini-inspection of the school, the Government Inspector had drawn attention to "inappropriate touching" as a teacher gave a piggyback ride to a small child. When asked what touching *was* appropriate, the Inspector's answer was unequivocal: "no touching." What we were doing as researchers felt like another case of "inappropriate touching," and asking questions about touching in Summerhill felt contaminating. Purity and prurience kept changing place.

There is a contrast between the practical management of risk of whatever kind (most likely to be nuisance or petty theft in Summerhill situations) and the Inspector's anticipatory prohibition of the very possibility of error or even mistaken perception, always conceived as implicitly sexual in nature (that is, "no touching"). It was interesting that when we presented those sorts of "outside world" scares (for example, no adult and child together in an otherwise solitary situation) and explained the rationale to older Summerhill pupils, they exclaimed: "if they don't trust in the teacher to actually be on their own with pupils then that's just pathetic." Yet the staff nevertheless receive obligatory "child protection" training and inspection days, and consider some of it irrelevant to what goes on in Summerhill (for example, "What would you do if a parent came up to school smelling of alcohol?"—in a boarding school, not a very relevant example, as a teacher observed).

Another major difference is that pupils propose and police laws to ensure proper running of the school, their own privacy, and the rights of individuals. These laws are decided democratically, each child and teacher having one vote. They address specific problems as they arise, rather than anticipating problems in terms of universal prescription. Even the School

Laws have numerous specific exceptions. For example, Law 48: "Freddy can have a stick bigger than him."[3] In Summerhill, "touch" is not a sensitive issue—and indeed appeared a ridiculous intrusion when we brought it up. And, having brought it up, we became unwilling agents of the same sexualized culture of "risk" that we were committed to investigate impartially, even though there was no impartiality out there for us to adopt as a stance. As outsiders, we felt contaminating, but that was part of the data: we were subjects in transit across Summerhillian boundaries that we could begin to feel in terms of their difference, but not yet understand.

Summerhill Outside-In

Summerhill is an almost perfect panopticon, incapable of secrets: "There are no secrets here, they all come out. The children believe that there are *some*, but actually there are no secrets" (houseparent), and "there are no secrets, so it [abuse] couldn't happen . . . everything goes to the Meeting and is spoken about and sorted out" (teacher). It is a total institution with boundaries both invisible and powerful—school pupils/"downtowners,"[4] locals/Summerhill cosmopolitans,[5] limits to parental visits,[6] and hidden definitions of normal/abnormal. As a pupil informed us: "there's other people who have been teaching in *normal* schools but they find it a really big jump but they do in the end settle in just fine, and are just *normal*" (pupil, our stresses). No one is locked in or out, but the borders are not often crossed. Staff members are seldom off duty in reality; there is a problem of "getting time on your own." Pupils visit them at will: "sometimes you just randomly turn up and say 'hi' and have a cup of hot chocolate or something" (pupil). So without intending anything negative, we suggest that Summerhill is a very precise and reliable mechanism for the social manufacture of selves (see below).

The Meeting scrutinizes breaches of the democratically agreed Laws and legislates for and against transgressors. All adults and children are equally entitled to participate in discussion, criticism, and voting. On the occasions we attended the Meeting, around two-thirds of the pupils were present. At that time, there were extant 174 laws that had been voted in over time, including individual and minute prescription of behavior: for example, "Len [aged 5] can have a lighter that doesn't light but sparkles." The Meeting has the power to make law, and indeed to abolish any or all laws. As we watched in March 2005, an eleven-year-old sought permission to light fires in the woods. Law 79 says she is too young: "Only Shack and over are allowed matches and lighters." Her claim was that "I'm good with fires; it's a nuisance to find someone who's older." The meeting decided that she was responsible enough, but had to undertake that she would not light fires for others of her age or younger and/or leave them in charge. Even those deemed less responsible in that particular ruling voted in her favor. Each law is the product of debate and voting, and it can be unmade at any time. This is self-regulation with a vengeance. Every alleged transgression is considered in its own right. In each case, and in the Summerhillian's accumulation of cases over time, there are questions of right and wrong, serious or piddling, appropriate or inappropriate. In ways such as these, the school—though it is a community more than a school in the conventional sense—invites and receives an all-embracing allegiance from its membership, which is unheard of in state schooling. It manufactures Summerhillians whose loyalties may transcend those of country and home: "my life is more kind of here and not at home . . . I would call this [Summerhill] home, instead of back in [country]" (pupil aged nine).

When "being yourself" and "having your life" proves problematic, the Meeting not only makes laws but is there to advise and adjudicate. Disputes or complaints may be dealt with informally, or by Ombudsmen, who are older students of either gender appointed to be a first-point of assistance. Unresolved or serious complaints may lead an individual to "bring up" whoever has offended them at the Meeting. Any child or adult, in any combination, may do so. An eleven-year-old girl explains: "The point about the meeting is to make . . . *me* feel [our stress] that it was totally wrong, this is a strong warning. But if you do it again, we will fine you. If you make contact [violence] you won't get a strong warning, they will probably fine you some odd random things like 'Bully's List,'[7] no television, no screens, no social games."

But the real force of the sanctions is social rather than financial. The Meeting teaches the antisocial that they: "can't get away with this stuff because everyone thinks *I'm* a right twit now and *I* [our stresses] have to calm down and build relationships . . . the more they go to the Meeting [and are 'brought up'] the more fed up and vocal the Meeting gets . . . so it [the problem] does turn itself around" (pupil). There is a clear element of persuasion, and also of public shaming, in these arrangements, but no signs of scapegoating. It is held, even, that those "brought up" seldom resent their accusers, although we did come across a pupil who felt that was not the case for her. Basically, the Meeting disciplines by instilling a sense of right and wrong in pupils by a practical, case-based approach. Through repetition, the more general moral development of a sense of fairness and responsibility is developed. The internalization can be vivid: "you really have to use your head and think 'oh, can *I* do this?' like if you were about to carve *your* name in a wall, you'd think, 'oh, do *I* think that *I* can do this, no, *I* probably can't'"(our emphasis). It seems to us significant that

Summerhillians shift between "I," "you," "we," and "them" in the way that they do. The speaker above hypothetically incriminates herself as an "I," envisaging herself considering an infringement. Summerhillians sometimes also referred to the "I" as a first name, "Katie," "Vicki": "I wasn't the real Vicki when I was in state school, you're not yourself." Transgressors were seldom discussed in terms as simple as an impersonal "he," "she" or "them." This kind of consideration was extended to their treatment of their interlocutors. It is not often that a schoolchild says to a researcher, "Have you had a good experience so far"?

Far from the "free" image with which we started this account, Summerhill school has invisible boundaries, powerful inspections, binding agreements, and redemptive rituals, as well as a set of public punishments that prompt and enact acceptable ways to live together. These all act as an "outside-in" pressure that frames and disciplines interactions while developing identities and relationships, yet always with the possibility of change or resistance. We have suggested something of the "total" nature of pupils' engagement with these structures. Summerhill is a powerful mechanism, generating discipline from within, and without the coercive relations of a "normal" school. The school orchestrates a vortex of engagements from which there is no "backing away," as one pupil put it.

Summerhill Inside-Out

At the same time, the "school" has weak boundaries where conventional "schools" have strong ones. There are weak boundaries between different pupil age groups: "In Summerhill a five-year-old could be best friends with a sixteen-year-old and that would not be a problem" (pupil), and "every time a new kid comes you help them more and more and feel better about yourself" (older pupil). Similarly, boundaries between staff and pupils are minimal in comparison with a conventional school. Pupils visit staff in their caravans, "it's just like visiting a friend in their room." They do so informally and indeed can be shooed away in like manner. Within the school, there are also many weak distinctions between public and private spaces: "[sometimes] I tell them I need time on my own and they're quite good about that because in a similar way they wouldn't want me to hang about their rooms all the time" (teacher).

Another weak boundary is spatial. The classrooms are inside but the outside woodland is accepted as an equally important learning area—how to play, to make things. Pupils come and go as they will, unlike movements "downtown." Inside or outside, whatever the season, male and female pupils often dress in similar clothes, big jumpers and loose trousers (a phenomenon noted by other observers more than forty years ago).

The strong boundaries are for outsiders like us, who can't climb the trees, can't go to the bedrooms, or upstairs in the House. We need a vote of the Meeting even to attend. There is also a reciprocity theme here. Weak boundaries are places of negotiation rather than prohibition or permission and so Summerhillians are good at reading each other, as well as being experts on "themselves." This, as we will see, is integral to the practices and interpretations of "touch" at the school: "Each teacher is different. A different person takes a different amount of time to settle in, the same as students," (pupil) and "for everyone it's different, and as long as everyone is respectful and isn't interfering with someone else's space and how they feel about that, there's no problem" (pupil).

How do the participants make overall sense of this interlaced world of weak and strong boundaries? The pupils' metaphors of association centre on notions of "home," "family," "brother," and "sister," with the teachers most often portrayed as "friends really," "like visiting a friend." Asked what the weirdest thing about Summerhill was, one of the oldest pupils replied: "I think it's that we all get on." The adults noted the "enormous attachment" pupils had to the place; "it's astonishing really," but were more likely to refer to the "tribe" or "community" of Summerhill, a "community based on the rights of the child with some constraints about ownership and about property and things like that . . . a community based on friendship rather than a family based on friendship" (teacher). Both staff and pupils pointed to a central value of "trusting people," with pupils more often claiming that the relation is an equal one—"we're all like equal" (pupil), while some of the adults shied away from the "family" images, and noted that the relation was pretty equal, but not entirely: "I'm not sure it would be such a good thing if they knew as much about me as I know about them" (teacher), and "so in spite of anything that leads towards equality there is definitely a distinction between an adult and a child and I think any of the adults here would have to acknowledge that" (teacher).

Those qualities of trust, equal rights, responsibilities, commitment, honesty, and confidence in pro-Summerhill accounts can take on a "*Swallows and Amazons*" romantic flavor, and we would prefer to stress that these are not so much the *qualities* of the inmates or the community as the *work* of the school in the construction, and reconstruction, of selves. For example, a teacher stressed that: "the kids are confident with adults, they're not coy. It's [teaching at Summerhill] not for someone who's a prima donna. I don't care if a kid says 'fuck off'; I don't care what the kids say to me actually, I think the honesty of it all is very good." There is a *dispassion* within Summerhill, as well as a passion for it. As a teacher put it (puzzling us during the interview), "the kids are completely *neutral* about what they are doing" (our stress). We now interpret this in terms of that "dispassion" we noted above. That is, it isn't personal; there is a system that delivers consequences for actions. Your friend may "bring you up" but it is the Meeting that delivers judgment. The Meeting is not *them*, it is *us*. It's connected to the phenomenon of the "floating pronouns" that we noted earlier—the grammar of empathy, as it were. We began to see that theme as a thread through our numerous conversations with pupils and teachers. The Meeting was clearly central to the "neutral" functioning of the school as a learning experience for the pupils, as well as a demonstration of "equality" in relation to adult as opposed to child power. As one teacher commented: "it's good for the kids to see you can 'bring up' adults," while another commented on the confidence that even little kids showed in the Meeting:

> Everybody is really quiet to be able to understand them so they really get heard and I think for the little kids that must be amazing to see these much bigger people all listening to what *I'm* [our stress] saying, it must be an amazing confident feeling.

One of the oldest girls commented that Summerhill was "a different way of life and a different way of education." We were struck by that coincidence of "way of life" and "education" in the social mechanisms of the school. The notion of "living your own life" was a dominant aspect of the culture. There is a pattern of social learning within little groups, often of different ages, which was very apparent in the data: "I always tend to have friends who are older than me because when I mix with kids my own age I don't learn anything, whereas if I mix with kids younger, I teach, er, give skills to the younger kids and the older kids give skills to me. So it's a win-win situation" (female pupil). The avoidance of the word "teach" may be significant. It's a little too directive for the culture. In fact, to understand "education" at Summerhill you have to be ungrammatical: the pupils *learn each other*, in more than one sense. And the teachers are *part* of that.

Summerhill also appeared to staff and pupils as a place of necessary risk. The grounds were open to the pupils, tree-climbing was permitted, and—to pick out the feature that would probably most alarm the "risk culturalists"—older children were allowed to carry "machetes," defined by Summerhill Law as blades over six inches (Law 94), but "Law 80: No sheath knives downtown (UK law)." Note the strength of the boundary implicit in Law 80. The "outside" is the UK, and by implication Summerhill is somewhere else—another country seemingly. Most concerns about safety were not about sexual threats of any kind; they concerned injuries caused by play, mainly skateboarding. Pupils were adamant about the value of "risks" such as swinging from the Big Beech tree: "if you didn't do that sort of thing you'd never have the chance to grow up," and "whatever you do there's a chance you'll hurt yourself and if you can't have chances like that, you can't live."

Finally, it is important to link these experiences to perceptions of the "outside." Many Summerhillians are familiar with conventional schools. Indeed, for quite a few, it is their "failure" at these schools that has taken them there. A dominant theme is that adults in such institutions "distance" themselves from the children: "I find it very strange when I see people distancing themselves away from other people"; "all the teachers there [state school] stayed very distant . . . and didn't give hugs" (Kent, southeast England), "the teachers used to stay in staffrooms and kids stayed out of staffrooms in my school" (northeast England), and "when I came here it was such a relief. I felt like there was this weight off my back. I didn't have to go to the state school any more. I didn't have to be bullied for the rest of my life and I didn't have to pretend to be something I'm not" (pupil).

Another major theme concerned their sense of themselves, referred to previously, and the emotional ethos of the school they had attended, including issues of bullying and harassment. Summerhill, perhaps above all else, was somewhere where you did not have to pretend: "you can just totally be yourself and don't have to act or try to get people to like you because if you're yourself and they harass you or make fun of you, you can bring them up in the Meeting" (pupil). They noted the absence of sexual harassment and name-calling at Summerhill, contrasting that with their earlier experiences: "If someone of the same sex gives each other a hug there, they'd get harassed loads and thought to be gay or lesbian," and "if you're a little bit different then you'll be classed as a freak and they won't go near you. But here, it's OK, they don't care" (pupil). Again, that same, *neutral* "don't care."

Summerhill: Touching on Inside-Out and Outside-In

We soon discovered "touch" to be the wrong focus; it became apparent that "touch" made no sense without locating it within the culture of relationships that constitutes Summerhill's production of selves. In terms of our research focus, we had to retreat from "touch" in order to locate it in a more contextual way. *Relational touch* contained physical touch, in ways foreign to "outside" institutions where an accountability and risk culture determines touching practices in more rigid ways. In Summerhill, relationships were constituted by both the "outside-in" and the "inside-out" features of the culture. It was a question of the nature of the flows between these two surfaces. We prefer to call Summerhill a "culture" rather than an "organization" because it effected itself in more tribe-like ways. It *governed* itself, and in so doing produced a distinctive citizenship, one that we find difficult to name happily as "pupils" or "students" or "kids," given the range in ages (4–16) and stages of development. That active citizenship in turn was the generator of identity. It was active and self-formative, in that participants *chose* what to take an interest in, and in choosing learned something of their own desires, responsibilities, and identity. As a Summerhill teacher from North America explained it to us: "if you're a child you expect some guidance, but the basic raw thing is: do you or do you not have the right to choose what you do from morning to evening, to stand or fall on the choices. And here you *do*."

The outside-in boundaries of Summerhill constructed the features that we earlier noted as apparently and deceptively "dark"—panopticon, total institution, self-regulation, and surveillance, all of which made Summerhill a strong as well as a benign society. It is also a highly intuitive and tacit one. The principal spoke of Summerhill's community as a: "family or a tribe, I think it's like a tribe, but it's more than that, it's just a life area; it's an area where everything happens and it's definitely not a school" (principal). But at the same time, these strong and bounded relations were interlinked with, and helped generate, weak boundaries between age cohorts, learning spaces, and across teacher–child relationships. The strong boundaries ensured things like social and personal identity, safe spaces, effective government, and social redress. At the same time they enabled weak boundaries that provoked relationships based on self-knowledge and negotiated spaces that were potentially learning-rich in all sorts of social ways. People learned to read each other, and hence themselves, in a kind of social dialectic: in such interaction varying degrees of "relational touch" were negotiated. And the panopticon features were available, more or less, to all.[8] Of course, the opposition of "strong" and "weak" is inadequate in itself, because the "freedoms" of Summerhill could also be breached in the strong sense—that's what all the laws and Meetings were for. Such breaches, however, were part of how the school worked as an organism; they were how people learned. The Meeting was a place of conflict just as much as it was of consensus. The Meeting has been portrayed in utopian terms, but it would be more useful perhaps to see it as a "working dystopia," as part of the "organic moving space" (principal) of the community. It is maybe not too much of a paradox to say that one way the school worked was by breaking down and mending itself, rendering problematic social relations *explicit* as a specific, moral, emotional, and rational curriculum for communal and personal living as well as learning. Issues central to "relational touch," then, were an inherent part of these disputes.

In addition, these processes were fed by a series of informal learning sets, based on a myriad of relationships—teacher-pupil, pupil-peers, mixed-age groupings, and so on. This was Summerhill as a learning-swarm.[9] In most organizations or institutions, strong links mean constraint and coercion, but in Summerhill the strongly bounded features—like the Meeting, or the social circumferences of the school, the school as community—created spaces for people to feel that they could "be themselves," "live their own lives," "recover" themselves from damaging earlier experiences, live without "harassment," and successfully seek redress for whatever injuries befell them in the school itself. In that latter sense, the "outside" was also at the core of Summerhill, as a set of learning and living experiences that pupils had to work on—hence the many comparisons by pupils of various "state school" pathologies with conditions in Summerhill. Hence too, in a weaker sense, the same core/periphery relation existed for the teachers at the school. Summerhill dealt with the real world outside as well as inside, constantly turning the inside out and the outside in.

Our conclusion was that the School enabled its pupils to be Summerhillians, and to call that "themselves." Each sought, in an oft-recurring sentiment, to "get on with their lives." An experienced teacher at the school, familiar with free schools internationally, commented that there is "an accepted individuality and agency that I have not experienced happening anywhere else; a very, very definite personal narrative." We see this vividly in the data: "if you don't act like yourself, you can't get true friends," so "you're just yourself" (pupil). That was both the autonomy and conformity of the school. In such a relational circulation, policed in such a "neutral" way, physical "touch" was neither here nor there. You do it if you feel like it, and if not, you don't. Another stereotype bites the dust—Summerhill was some way from being the "touchy-feely" school it is sometimes portrayed to be.

"Lessons" for Educators

Although we started with the notion of Summerhill as a free school, locally known as the "do-as-you-please" school,[10] we found Summerhill to be structured in ways that were almost always neglected by Inspectors, media accounts, and academic comment (Chamberlin, 1989).[11] It will be recalled that there were 174 laws extant at the time of our fieldwork. But the structures enable rather than disable in the ways that might be anticipated by Foucault. This is a *benign panopticon*, within which various forms of learning are promoted as a result of the weak boundaries between staff/students, and also age cohorts in the community. We could consider social learning here (Lave and Wenger, 1990), or scaffolding (Vygotsky, 1978), connecting these to the informal learning sets that we found in the community. But these narrower theories of *learning* seem to be of less interest than theories of *becoming* in the Summerhill context. We need to understand how the plural and paradoxical panopticon works for the good, or at least for certain kinds of good, and how that connects with the notion of "relational touch." We concluded that the efficacy is not psychological so much as it is anthropological, much as Neill later argued (1971). Our theories of social organizations are helpful only if we invert them in this case, and see how "becoming" develops in the "being" of the Summerhill machine.

First, as we saw earlier, the notion of "touch" as an issue of "safety" or "protection" was regarded as absurd at Summerhill. Instead we developed the concept of "relational touch," wherein Summerhillians learned to relate to themselves, to others, and to intuit boundaries. The invocation of this kind of citizenship is what A. S. Neill had in mind. As we saw from the "floating pronouns" in their talk about the rights and wrongs of "cases," Summerhillians were culturally adept both at putting themselves in other minds and, more importantly, putting other minds in themselves. This is less like the liberal expression of difference ("they are just like us") than a more radical insight ("we are just like them"). The same distinction turns up in contemporary theorizing about the "projective imagination" (Tanesini, 2001, 18), and the possibilities of a notion of "we" that does not depend so remorselessly on an exclusive notion of "them." Much of the current debate on "voice" and "empowerment" in schools might do well to reexamine its insubstantial nature in contrast with the practices of Summerhill (Osler and Starkey, 2005).

Second, we also noted how democratic mechanisms within the school offered a visibility of practices that was far more "effective" than any conceivable transparency of procedures. The Meeting filled the gap between what Law and Mol (2002) call "managerialist" control (as it were, the very limits of audit) and the always excessive flow of complex, embodied interactions that characterize any organization. The prospective or retrospective rhetorics of audit were merely "staging accountability" (Law and Mol, 2002, 101)—as prevention or blame—while the agonistic realities of resolving real conflicts and injustices enacted responsibility *in the moment* and *for the moment*. Of course, this practical/procedural dichotomy was breached continuously in Summerhill by the creation of new laws and adjudications in respect to members. But these latter regulations were part of the practical flux, open to adaptation, extension, or repeal. They were not a universal template for measuring compliance so much as a situated and shifting search for resolution that regarded its short-term failures as ultimately productive: they approximated self-government, not governmentality. In relation to Law and Mol's argument, the audit solution, advocated by HMI, amounts to a form of "utopian absolutism" (regulations will prescribe and proscribe actions in order to achieve Best Practice), while Summerhill took a more pragmatic and ad hoc approach. In this way, conventional attributions of the utopian and the pragmatic change place.

More generally, such an approach offers a corrective to the "Risk Society," which many claim is responsible for polluting touching and other behaviors (Beck, 1987, 1992). Elsewhere, the "risks" are managed not by managing and distributing the "goods," but by managing and distributing the "bads"; performance is very much focused on danger. To invert Law and Mol's (2002) taxonomy of utopianism, this is "dystopian absolutism." However, such accounts offer only a partial reading of "expert–lay" relations and "fail to recognize that the 'done to' lay public are at one and the same time 'doers' working within the relations of definition" (Mythen, 2004). In other words, both professionals and lay public share an experience of helping to create particular labels and definitions that in turn provide sets of relationships that derive from them. In a situation where the "non-risky" population now view themselves at equal risk with the "risky" population, an element of self-fulfilling prophesy begins to circulate.

Third, in addition, Summerhill embodies democratic practices that work in relation to the development of pupil identity and democratic, friendly relations between pupils of different age, gender, ethnic group, and nationality. That is hugely important in that, elsewhere, these things don't happen. Bullying, harassment, racism, and alienation are rife in schools. Summerhill outcomes realize the rhetorics of the Universal Declaration of Human Rights: Article 22 promotes practices "indispensable for their dignity and the free development of their personality" (cited in Dews, 2002, 36). Such concerns are peripheral to the dominant audit regime, with its insistence on standards, effectiveness, and improvement. But they have something important to say to the policy maker stuck in the narrow instrumentalism and authoritarianism that characterizes contemporary schooling. The current accountability monologue is damagingly antieducational in this respect, and it is interesting how progressive notions have disappeared from public discourse. It was once suggested by Inspectors that Summerhill constituted a "piece of fascinating educational research" (HMI Report, 1949, cited by Goodman in Lawson, 1973). Michael Young even "suggested that the progressive schools might become laboratories for educational research and experiment, attached to university departments of education" (Skidelsky, 1969, 256). That research future was never realized. It may seem a strange historical anomaly that we need to turn once more to the legacy of "free" schooling, but the recurrence is even greater. There is much in Dewey's writing on the school as a "miniature community, an embryonic society" (Dewey, 1962, 18), as a place for the development of active beings whose "interest" is in education (Dewey, 1966, 125). Indeed his definition of the notion of democracy itself is a good description of Summerhill in educative practice: "A democracy is more than a form of government, it is primarily a mode of associated living, of conjoint communicated experience" (Dewey, 1966, 125).

Fourth, there is a corollary to those damaging circumstances. The radical separation of educational research from educational experiment left "change" in the hands of the politicians and the media as the most potent evaluative voices. It left progressive schooling as a backwater, in policy terms. It left educational research bereft of experiments, and so increasingly tied to evaluating the "innovations" dreamed up by politicians or "outlier" researchers willing to advance government causes in education. In that sort of way, Summerhill simply "disappeared" from the mid-1970s until 1999, when OfSTED decided to deliver the *coup de grace* and try to close down the school. The consequences are too numerous to go into in this chapter, but two of them should perhaps be noted. First, we found it difficult to interpret Summerhill as an educational entity—it did not fit into the improvement and effectiveness discourses of educational research and evaluation. It did not fit the audit templates of OfSTED. We did not find sociology of education theories that seemed to address its unusual nature. We therefore adopted a "grounded" approach to understanding the nature of the case, and turned Theory upside down in order to better match the distributions of power in the case—"benign panopticon" and so on. There is a second possibility. Maybe we do not have good theories of schools because we do not have good schools to theorize about.[12]

Fifth, and finally, can we disentangle some of the political context? It is salutary to return to the literature of the 1960s. Was there ever really a time when a skeptic of progressivism, not an advocate, would write of education in the UK: "No one would dispute the claim that the progressive ideal has triumphed, or is triumphing, at the primary and junior level" (Skidelsky, 1969, 14)? Or was there ever a time when a major educational figure in the United States[13] would find in the self-government of Summerhill "the breakdown of our Western code of morality implicit in the spread of Neill's hedonism to the majority of the next generation" (Rafferty, in Lawson, 1973, 20). The same critic ranted his way through images of "sex perverts" (15), "frolic in the park, a daisy-picking foray, or an experiment in free love" (17), concluding: "[w]hat the unkempt and sometimes terrifying generation of tomorrow quite obviously needs are more inhibitions, not fewer" (18). These were the surfaces of some of the moral panics that Summerhill engendered. But there were opposing verdicts from the other side, black radicals in the United States who saw in Summerhill an end to oppressive, state-administered schooling. Michael Rossman saw Neill as a

Spock for "the 'post-modern' young just now maturing into parenthood" (in Lawson, 1973, 141). Nathan Ackerman exclaimed: "Shall a child be governed from the head down or from the heart up?" (242), while Erich Fromm saw in Summerhill a stark dichotomy between, "love of life" or "biophilia" and the current "necrophilia" (253). Little wonder, on any side, that Summerhill became a shifting signifier, demonized on the Right and endowed with magical properties for social revolution on the Left.

Certainly, the sorts of progressivism and critical education that Summerhill and its offshoots and developments stood for—or were taken to stand for— were buried by the Thatcherite resettlement of the 1980s. In the United States, Giroux claimed that "[a]t all levels of national and daily life, the breadth and depth of democratic relations are being rolled back" (Giroux, 1992, 12). In the UK these trends have culminated in the sorts of authoritarian micromanagement of educational acts that "touch" regulations exemplify, and also in the precise specification of teaching performances within a more general regime of National Instruction. Such logics have reached a kind of absurd intensity that in itself must engender a countermovement. We end by returning to Paul Goodman, who commented on Summerhill and progressive contexts in the early 1970s. He argued that just as Rousseau opposed the artificiality of monarchy, and Dewey the genteel residues that were irrelevant to industry, so too had Neill reacted against twentieth-century authoritarianism. Neill had in mind, of course, not "60s permissiveness" (where he is usually and deliberately mislocated), but the European moral abyss that was exemplified by the Great War. He then had in mind the authoritarianism of Fascism, and had the privilege of being refused a visa first by the Soviet Union and then by the United States in its McCarthyite phase. Goodman concluded with a generalization that we offer as an optimistic prompt to further thinking on progressive and critical educational reform:

> The form that progressive education takes in each era is prophetic of the next social revolution. (Goodman, in Lawson, 1973, 213)

Notes

1. Economic and Social Research Council (UK) grant number RES000220815. "Touchlines: The Problematics of 'Touching' Between Children and Professionals." (Piper, H. Stronach, I. and MacLure, M.)
2. We are very grateful to all the participants in the Summerhill community. This account has been negotiated with the Principal of the school.
3. This rule of course implies that current law dictates that it is usual to have a stick up to one's own height.
4. This is the term used to describe people in the local village.
5. The school intake is international—the staff is mainly but not exclusively British.

6. The principal had no concrete limit, but visits were usually "about twice a term" on agreed days. Otherwise, "it disrupts the whole life of the community to have people coming and going."
7. An adult explained elsewhere: "the Bully's List... doesn't mean that you are a bully, it's just the harsh fine." There were no pupils on Bully's List at the time of the fieldwork.
8. The classic panopticon disciplines by making the mass visible to the master (Foucault, 1977). But in Summerhill, the term can be radically distributed, in that each member has a perspective on the others, which they hold to be complete, or almost so. Hence the repetition of the "no secrets here" theme. It is clear that perspectives on the other are unusually mutual, unmediated, and visible in the Summerhill community.
9. It was this aspect of the school that the Government Inspectors most consistently neglected. In the last full inspection of the school (1999) by HMI/OfSTED, only one data-recording sheet (out of fifty-four lodged by the Inspectors) addressed learning outside the classroom. Inspectors regarded what happened outside the classroom basically as a kind of truancy, hence their obsession with the question "how often do you attend lessons"?
10. The "do-as-you-please" tag is current. It is long-standing (Skidelsky, 1969, 33). In its inaccuracy, it was useful for justifying the criticism of Summerhill as "narcissistic" and inevitably "individualist" (Tam, 1998, 57).
11. Chamberlin's philosophical account of education and freedom takes individual autonomy at Summerhill too literally. Accordingly she counts Neill an "extreme" libertarian, and offers the usual disclaimer: "The freedom to choose what line of study to pursue and how best to pursue it is inappropriate for children whose intellectual skills are relatively underdeveloped" (1989, 110).
12. The discourses of accountability and audit would certainly deny this. In relation to schools' "excellence," "quality," and "effectiveness," verdicts are abundant, worldwide. But they rest on narrow performances of "schooling" and often fail even to address questions of value.
13. Rafferty was in charge of the Californian school education system. He makes OfSTED's HMCI at the time of the attempted closure of Summerhill seem measured in his views.

Further Sources of Information
(Particularly useful to readers without prior knowledge of Summerhill)

Ayers, W. (2003) *On the Side of the School. Summerhill Revisited*. New York: Teachers College Press.
Lawson M. (ed) (1973) *Summerhill: For and Against. Assessments of A.S. Neill*. Sydney: Angus and Robertson.
Neill, A.S. (1937) *That Dreadful School*. London: Herbert Jenkins.
Neill, A.S. (1939) *The Problem Teacher*. London: Herbert Jenkins Limited.
Neill, A.S. (1968) *Summerhill*. Harmondsworth: Penguin (1st pub. 1962).
Neill, A.S. (1971) *Talking of Summerhill*. London: Gollancz.
Vaughan, M. (ed.) (2006) *Summerhill and AS Neill*. London: Open University Press.

Chapter 4

Quaker Schools in England: Offering a Vision of an Alternative Society

Helen Johnson

Who are the Quakers?

A straw poll about Quakers taken in a university cafeteria is likely to bring a variety of responses. Replies are likely to be prefaced with some form of qualification that admits a lack of detailed knowledge but some impressions tend to come up time and again, something about pacifism, social responsibility, and the invariable joke about porridge. There will be a mixture of impressions about beliefs and behavior in society, and dietary habits, a plainness of appearance, and, more deeply, a seriousness and thoughtfulness in lifestyle. (The official Web site, *Quakers in Britain*, however, makes it clear that Quakers are not as serious as they sometimes appear, that most do not wear black all the time and some do not even eat porridge every day.) However, even among university coffee-drinkers who have a built-in propensity to critique, the sincerity of this very small religious group is rarely, if ever, questioned. However, to understand the position and role of Quaker schools in England, it is necessary to have an awareness of the reality of Quakerism today. (Interestingly, while recently there are perhaps signs of membership growth, Burnet (2007) has written about what he terms the failure of Quakerism in Scotland). The information presented here has been collected through a review of the literature, documentary analysis, an analysis of school prospectuses, and interviews with those working in the field of Quaker education.

Questions, Equality and Tolerance rather than Dogma

Quakers have never easily accepted conventional wisdom and, at their origin, can be seen to be part of the social upheaval in England of the seventeenth century. Their founder, George Fox (1624–1691) was a man who asked questions of the prevailing social and political mores and in so doing was seen to be undermining—or at least, challenging—the authority of the established Church of England and its concomitant power structures. For he placed great emphasis on the uniqueness of the individual, and was supported by a group of those who were like-minded. Fox's views about the possibility for an individual to have a direct relationship with God (or a higher power) without a priest as an intermediary can be seen to place him and his followers in the vanguard of Protestantism and so, in that context, be social and religious revolutionaries. MacCulloch (2003, 526) in his mammoth history of the Reformation, summarizes the Quakers' historical evolution in these terms:

> ... some gathered around the powerful figure of George Fox, who was determined to harness ... energies to a godly purpose and end the excesses. Those who (more or less) accepted this centralizing direction came to call themselves "Friends of the Truth".... At the time the Friends horrified and infuriated everyone else by their deliberate flouting of social convention to show that they acknowledged no power but God's: they refused to doff their hats respectfully to social superiors....

MacCulloch (2003, 526) goes on to point out that almost alone among the rebellious groups of this time, the Quakers:

> ... survived and transformed their character into a peaceable, self-restrained people... Their present-day association with peace, disarmament and ecological movements is a quiet return to their original commitment to questioning all established authority.

Then and now, rituals and ceremonies were unneeded—an attentive silence was the means to spiritual awareness and understanding–and the presence of God was to be acted out in the everyday lives of ordinary working people, be they men, women, or children. Of course, the meetings for worship were and are not necessarily entirely silent. Then and now, individuals may feel moved to speak and since that inspiration was regarded as coming from the indwelling Spirit of God, the gender of the speaker was of no concern. In this way, women achieved parity with men in a spiritual and organizational sense (see Levenduski's, 1996, exploration of female spirituality). Significantly, given the maleness of hierarchy, the Quakers have a bare minimum of formal hierarchy (for example, Monthly, Quarterly, and Yearly Meetings held to conduct the Society's business). From the seventeenth century, Quaker culture has evolved into the egalitarianism, tolerance, and a bottom-up organizational democracy (albeit that decisions "emerge" and are not put to the vote) that it embodies and promotes today. But this leads to other issues, especially in the transmission of this culture to succeeding generations. In this, the difficulty with the Quakers is, of course, their *lack* of dogma. No fundamentalist zealots are they (West, 1962) with rigid—and so easily "teachable"—"rights" and "wrongs." Gillman (2003, 7), in the introductory book sent out from Friends House, London, to all enquirers about Quakerism, says it plainly: "You will not find the convictions of Quakers in this country set out in any creed." Quakers have always tried to live out their faith as in George Fox's famous words: Let your life speak! An outward conformity to a creed is not enough. The *Quaker Faith and Practice* (1994, revised 2005, 17) says succinctly: "In the Religious Society of Friends we commit ourselves not to words but to a way." Confirming that Quakers have a "certain mistrust of theological terminology," Gillman (2003, 7–8) makes the point "that words are often divisive." After all, he points out their epistemological uncertainties:

> The same word often means different things to different people and similar ideas are sometimes expressed in very dissimilar ways.

Gillman goes on to quote William Penn (1726, 781), the founder of the state of Pennsylvania, who makes it clear that Quakers are concerned with *practice*, the living out of elevated or transcendent values:

> ... it is not opinion, or speculation, or notions of what is true ... that ... makes a man a true believer, or a true Christian. But the conformity of mind and practice to the will of God, in all holiness of conversation (= behaviour), according to the dictates of this divine principle of light and life in the soul.

However, in contrast to their lack of dogma, Quakers offer information about virtually every aspect of their practice and lifestyle in their publications (for example, *Quaker Faith and Practice*, 1994, revised 2005), school prospectuses, and Web sites. Deal and Kennedy (1982), in their now famous one-sentence summary of what is meant by organizational culture, talk of "the way things are done around here"; and the *Quakers in Britain* Web site certainly goes into detail. For example, in response to frequently asked questions, the relevant section tackles the issues head-on that were identified by the university coffee drinkers. While confirming that there is no dogma or doctrines that have to be adhered to, the Quaker Web site (www.quaker.org.uk) goes on to say:

> There are, however, some commonly held views which unite us. One accepted view is that there is that of God (or the spirit or divine) in all people and that each human being is of unique worth. This shared belief leads Quakers to value all people and to oppose anything that harms or threatens them.

It is to this common core of values that we now turn.

The Common Core

Quakers in Hampshire (www.hampshirequakers.org.uk/quakers/testimonies) have recently listed testimonies "valued by Quakers around the world" (that is, personal relationships, compassion, lifestyle, truth, social justice, and green issues). Lists can vary, but the one always included in such a list is the peace testimony, the one that is most readily associated with them. Since 1660, Quakers have said "We utterly deny all outward wars and strife and fightings with outward weapons, for any end or under any pretence whatsoever, and this is our testimony to the whole world" (*Quakers in Britain*, 2007).

More recently, the question has been asked, even within the Society, whether or not Quakers are Christian. The official response makes it clear that:

> Quakerism started in England in the 1650s so there's no doubt that Quakerism is rooted in Christianity and many Quakers centre their faith on Jesus. On the other hand, some Quakers find that traditional religious language doesn't describe their inner experiences, and they look both within Christianity and within other faiths and philosophies. . . . (*Quakers in Britain*, 2007)

As mentioned earlier, an important document is *Quaker Faith and Practice*, published by the Yearly Meeting (the recent edition was revised in 1994, published in 1995, and reprinted with further revisions in 2005). It essentially sets the norms of the Quaker culture: for example, General Council on Church Affairs; Yearly Meeting; the Meeting for Sufferings; membership; Quaker marriage procedure; Quaker funerals and memorial meetings. Within the hefty volume is a section headed *Advice and Queries* that in the place of a creed offers particular advice and prompts for self-questioning and reflection; and so in this way, a critical epistemology, underpinned by the principles of freedom and diversity, is built into Quakerism. Members are counseled, for example, in the latest version, to:

- Try to live simply. A simple lifestyle freely chosen is a source of strength. Do not be persuaded into buying what you do not need or cannot afford. Do you keep yourself informed about the effects your style of living is having on the global economy and environment? (paragraph 41).
- Respect the wide diversity among us in our lives and relationships. Refrain from making prejudiced judgments about the life journeys of others. Do you foster the spirit of mutual understanding and forgiveness that our disciplineship asks of us? Remember that each one of us is unique, precious, a child of God. (paragraph 22).
- Remember your responsibilities as a citizen for the conduct of local, national, and international affairs. Do not shrink from the time and effort your involvement may demand. (paragraph 34).

An Outward-Looking-and-Acting Community

The Quaker historian, John Stephen Rowntree, estimated that in 1680 there was something in the region of forty to sixty thousand Quakers in England and Wales (Walvin, 1997, 92). Today, the Quakers are an even smaller religious group. Figures do vary, but there are approximately 17,000 Quakers in Britain today (Heron, 2000), with 400 Quaker meetings for worship each week. Additionally, it is estimated that around 9,000 people attend Quaker meetings regularly without being formal members of the Society. It is also estimated that there are about 210,000 Quakers worldwide (BBC, 2007). Quakers have a substantial presence in the United States where members fled to in the seventeenth and eighteenth centuries to escape religious persecution, and while American and British Quakerism have great commonalities, their different histories reflect their own particular social and religious contexts and structural differences and social impact (Yount, 2007).

Much has changed in their social and moral attitudes. Once puritanical, they are now very progressive and accepting of minorities. (In November 2006, readers of *The Pink Paper*, the British gay newspaper, nominated the Quakers as the most gay-friendly organization in Britain). However, as mentioned earlier, the Meeting for Worship with its silent stillness remains the cornerstone (Gorman, 1973), along with the peace testimony mentioned above, of Quakerism. Again, quoting the *Quakers in Britain* Web site (2007):

> Quakers think that everyone can have a direct relationship or "communion" with God. We find that this communion can best be experienced if we meet

in silence, with nothing pre-planned. Meeting for Worship couldn't be simpler: you go in and sit down in a plain room with other Quakers and settle into silence, a silence which can become very deep and powerful.

Could such a direct relationship with God make them elitist or self-regarding? Does such "plainness" separate them from society? Are they a small inward-looking group? Could they even be described as a sect? The quickest of looks at their history soon finds figures such as the prison reformer Elizabeth Fry (Heron, 1997), progressive employers such as the Cadburys and the Rowntrees, and Joseph Lancaster, who opened many schools in the early nineteenth century for poor children. There is much evidence that confirms their commitment to social action and justice. In response to George Fox's entreaties, today's Quakers are very much engaged in changing the world. They are active in the fields of

- peace and reconciliation
- world development and global citizenship
- homelessness
- prison reform
- international organizations, for example, Amnesty International and Oxfam
- education, where a particular concern is bullying in schools and conciliation between children
- enlightened employers
- ethical business.

So, far from being "exclusive," Quakers live out the interconnectedness of a core of elevated or transcendent values in the world. In a much quoted and influential Swarthmore lecture in 1967, Kathleen Slack argued for the Quakers to leave their position on the margins, talking in terms of an "open" and "closed" society:

> An open society is one in which all members have a liberty in marked contrast to its absence in a closed society, membership of which is, in one form or another conditional, and which preserves to some degree or in some way its identity as distinct from society as a whole . . . The open society has no such characteristics. In it, there are no agreed, absolute or unchanging ends or ideals. It is fixed and unfixed. (as quoted by Heron, 1997, 31)

Such a view has a fluidity that Bauman (2000) for one would recognize in the postmodern twenty-first century world.

Interestingly, for a group whose membership is now from predominantly the middle and professional classes, George Fox was a weaver. He was a man with a perceptive intellect but without a sophisticated education. But as the Quakers grew and developed, because of the lack of a priest-class, there was a need for a literate and educated membership. With this need came schools and, as early as 1668, day schools were founded for Quaker boys and, significantly, for girls, and these were soon opened to other children. This education was not simply about Christianity but was to include "whatsoever things were civil and useful in creation" (Quaker Faith and Practice, 1994, revised 2005, 23).

So, again it can be seen that the Quaker education was to be placed in a wider, civil context and to be useful and practical. While in many ways, Quakerism can be seen as a religious tradition in which "tennis is played without a net," it is clear that it does possess a recognizable culture. To substantiate this, reference has been made to both publications (text and virtual) and to the managing structures of Quakerism. Thus, a religious radicalism and freedom coexists with necessarily conservative organizational structures and procedures. Dandelion (1996) has written of a Quaker double-culture in which a culture with a liberal belief operates alongside a conservative and conformist culture. So it would appear that within the "openness" of mind there is something more complex than indicated by the Frequently Asked Questions and Answers section of the Quaker Web site (www.quaker.org.uk).

Why Quaker Education?

How does the simplicity and complexity of Quakerism translate into a school setting?

From the very beginning there have been certain aims and expectations. Dr. Fothergill founded Ackworth School as the first Quaker boarding school in Yorkshire in 1777. The founding philosophy of this school, based on a clear sense of the transcendental and practical, was:

> . . . to provide a tender teachable disposition . . . and make way for the softening influence of divine good-will in their hearts, filling them for the fruitful discharge of every duty in life. (as quoted in Walvin, 1997, 95)

As has been pointed out, the American experience of Quakerism has in many ways been different from the British one. However, there is substantial "crossover" in terms of values and approaches to education. In 1920, John Lester, the first executive director of the American Friends Council on Education made it clear that Quaker education was to be seen as "progressive" education. It was these "progressive" values revolving around peaceful resolution of conflict in the macrocosm and microcosm, equality, tolerance and simplicity in lifestyle that are consistent with Quaker values and are attractive to parents, Quaker and non-Quaker, who wanted their child to be exposed to such a moral climate and community. Lester (1920, 8–9) put it in these terms:

> There is the concern of the Quaker school to discover the means by which children of various age levels may experience a power outside themselves which makes for righteousness. The pervasive and basic purpose of the Quaker school in the conditioning of growth in children is not only to equip them to realize what they can do best, or what they will be most happy in doing. It is not even that they shall realize what work is most worthwhile for them to do, or what tasks most need immediately to be done for the welfare of their fellowmen. The pervasive and most basic purpose is to enable them to come to an understanding with themselves about what is right for them to do, and then to a desire, an urge, a resolve to do it . . .

In this way, Quaker education can be seen either formally or implicitly as being under the "progressive" education umbrella. As seen in contemporary "progressive" initiatives in Great Britain, such as Human Scale Education, there is a sharing of approach to holistic personal development and social action. In Dewey's terms (Soder, 1996), this means the growth of the engaged citizen and the development of a critically engaged intelligence.

In Great Britain, fifty years ago, Harold Loukes (1958, 7) asked pertinent questions about the transmission of values to children. Those parents who have a certainty about their own place in a secure and declared religious tradition "with a clear framework of belief and custom, which may be authoritatively expounded and progressively accepted and understood," have perhaps a simple path: to expose their children to such certainty. However, Quaker parents have no such certainty. Should they even expose their children to Quakerism? As Loukes (1958, 123) points out, Quakerism needs a particular kind of maturity, one that "lies in the ability to live without certainty." However, as he goes on, "A child needs to be sure." What is "right" and what is "wrong"? Or is a way forward to ensure that the child has some sense of boundary? That is a theological issue, but it is also a pedagogical one.

There are also important social and political issues. For, as Loukes (1958, 97) says, "Friends are committed in principle to an attack on privilege and social distinction, but find themselves possessed of schools within a privileged and class-conscious system." Significantly, the issues of class and social division that Loukes was writing about nearly fifty years ago in England have not gone away. In fact, in many ways they have become heightened. The argument for the seven English Quaker independent (boarding and day) schools (Ackworth, Saffron Walden, The Mount,

Bootham, Leighton Park, Sibford, and Sidcot) remains—that the schools are a particular site where a particular culture can be experienced. (There are no Quaker schools in Scotland and Wales; there is one in Northern Ireland and one in the Republic of Ireland). The *Quaker Schools* Web site (2007) states, with echoes of those "progressive" values again, that "in their attitude to education Quakers try to ensure that:

- every learner is approached hopefully,
- individuals are encouraged to believe in their own immense potentialities,
- learning happens most creatively when relationships are based on mutual respect,
- methods of discipline are based on trust and mutual support and a desire to promote the positive,
- responsibility is encouraged, as are questioning and exploration, honesty and openness."

In themselves, there is nothing here that rings of social elitism and privilege. In earlier editions of his introduction to Quakerism, Gillman (1988) can perhaps be seen to avoid the issue of social privilege. He noted that "The majority of Quaker teachers work in state schools . . ." He went on to state that in the independent boarding schools "the academic side of life is respected but it is the development of the whole personality which is the main goal. This includes the encouragement of artistic and practical skills." Once in the eighteenth century, the then numerous Quaker boarding and day schools were seen as "the key to the Society's future" (Walvin, 1997, 94). Today, the role of the seven fee-paying and independently run Quaker boarding schools (that also have day attenders) is far more controversial.

In *educational* terms, there is also a tension between these Quaker schools as a way for the child to be withdrawn from the world—to be made safe and secure in comfortable surroundings—and for that same child to be prepared for engagement with that same world. Meanwhile, the fierce debate among Quakers about their very existence has intensified. The Letters page of "The Friend," the weekly Quaker newsletter, frequently carries letters from readers defending or attacking their existence (see, for example, correspondence in editions published in the month of April 2005 and months of November/December 2006).

Today's Schools Speaking for Themselves

Much research into Quakerism does seem to be historical (Levenduski, 1996; Burnet, 2007; Yount, 2007) and revolves around particular historical figures such as Rowntree (Vernon, 1982), George Bishop (Feola, 1997), and Elizabeth Fry (Hatton, 2005). Recent research from a social sciences perspective seems less easy to come by, with some notable exceptions—for example, Pink Dandelion's (1996) *A Sociological Analysis of the Theology of the Quakers* and Nesbitt and Henderson's (2003) study of *Religious Organisations in the UK and Values Education Programmes for Schools*, which focused on the Brahma Kumaris Spiritual University, the devotees of Sathya Sai Baba, and Quakers.

Specifically, what do people who work in the Quaker schools say about what they do and why they do it in the way that they do? As noted briefly on the first page of this chapter, in this small research study reported here, the data were collected through a review of the literature, documentary analysis of the prospectuses of the seven English Quaker schools, visits to three of these schools, and a series of twenty interviews—with the head teacher of a Quaker school, administrators, Quaker teachers (working in both Quaker and state schools), and other Quakers with an interest in education (two of whom had attended Quaker schools as pupils). Also, what Quakers say about themselves in official publications, newsletters, magazines and journals, and, of increasing importance, Web sites was surveyed. It is important to note that each of the seven Quaker boarding schools is independent from both the state in terms of funding and from the Religious Society of Friends as a controlling and managing organization.

During the series of interviews (supported by the analysis of school and other documentation), anonymised Quakers active in the field of education offered these answers to relevant questions. The following reports their synthesized perceptions:

1. The defining features of Quaker education.
   - Very early in their history, these boarding schools were about indoctrination to educate (and assist the retention) of children of marriages where one partner was not a Quaker. Now they are open to all; in fact, most children who attend are not Quakers. Pupils of all religious beliefs are welcomed: Jews, Hindus, Buddhists, and Muslims, evangelical Christians and Roman Catholics and nonbelievers.
   - They are not owned or controlled by the Society of Friends; they have a close relationship with school governors from the general meeting of the local meeting/branch and espouse and promote declared Quaker values. Unlike some faith schools where it is possible to visit them and not be able to recognize their denomination, Quaker schools embrace—in their own way—Quaker values, and their value position is made clear in public documents such as the schools' prospectuses and the Web sites.

- Each day starts in a recognizably Quaker way with a silent gathering for worship. It sets the tone for the day as a meditative experience. For all those interviewed, the peace testimony, in which Quaker pacifism is the core element, is an important part of the schools' reality. There is a Quaker Schools Week when the children engage in projects looking at war and peace. The schools conform to the National Curriculum in England but recognize the uniqueness of every child. The schools emphasize living as a small community with a regard for democracy (for example, Sibford School has a student forum that offers the children experience and practice in democracy). The number of children with special educational needs (for example, according to the head teacher, 30 percent of Sibford pupils have dyslexia) varies among the seven schools. Other schools (for example, The Mount and Bootham, both in York) can be regarded as being academically high achieving.
- There is no "party line" taken by the schools, but the statement of aims of each school is discussed by the seven head teachers and the two head teachers of the Quaker schools in Northern Ireland and the Republic of Ireland at the Quaker Schools Heads Conference. So it is left to each head teacher to define "Quakerness" and how it will be demonstrated in schools where very few of the children, staff, and head teachers are themselves Quakers. The only head teacher of the seven who is a Quaker is the one at Sibford. He also chairs the Quaker Schools Heads Conference.
- The ethos and atmosphere of each school is important and overt statements are made about Quaker culture as demonstrated, for example, by the displays of school notice boards and a calm, supportive environment as observed by this researcher. (Such inferences are supported by reports of the Independent Schools Inspectorate in England. See, for example, the inspectors' report: "Leighton Park is highly successful in providing an education of good quality in a strongly Quaker ethos." (Independent Schools Inspectorate, 2004)).
- Parents are seen by the teaching and administrative staff of the schools as sending their children to a Quaker school (especially if they are not Quakers and may not even be churchgoers) because they want for their children morality without dogma. Again, parents are perceived by the school staff as wanting their children to have a spiritual dimension through the undogmatic means of the *Advice and Queries* (contained within the *Quaker Faith and Practice* volume, 1994, revised 2005).
- An unquantifiable number may be specifically attracted by the peace testimony. Interestingly, families in the armed services also

send their children to Quaker schools. The school administrator, who drew this phenomenon to the researcher's attention, could offer no conclusive reasons for what, on the face of it, seems a strange contradiction.
- All the teaching and administrative staff interviewed noted the current popularity of Quaker schools. Busy, middle class parents, in particular, were regarded as finding Quaker schools attractive as they supply a caring and activity-filled environment, as well as small class sizes (virtually half the state school average of 25–30) and the overall size of the schools (again almost half the size of the average state comprehensive schools, which have around a thousand pupils, while others can approach nearer two thousand). Bootham School, in York, is a typical size for a Quaker school in England and has approximately 430 pupils.

2. The schools' innovative ways of nurturing rounded personal development.
   - Quakers are urged by the *Advice and Queries* (to be found within the Quaker Faith and Practice volume, 1994, page 12 paragraph 27) to "live adventurously," and as has been seen above (for example, the comments of Loukes, 1958, and Lester, 1920), Quaker education does place an emphasis on values and creativity, rather than only focusing on the outcomes of a governmentally imposed "performativity" (Ball, 1995) that is seemingly focused on pass rates, league tables, and standard assessment test scores .
   - There is a variety of academic, personal, and social activities to encourage the holistic development of the young person. One school magazine, for example, displays the range of student life and activities: involvement with the National Academy for Gifted and Talented Youth, based at Warwick; charity work; Young Enterprise; a sports tour to Cyprus; foreign trips; domestic trips; open day; water sports; the Duke of Edinburgh Award; performance projects; environment/horticulture; art gallery; pottery studio; music; dyslexia department; and textiles.

3. Responding to what parents want.
   Spiritual, moral, cultural and social development, both through the curriculum and other activities, in any school is underpinned by the ethos and values of the school (DfEE, 1994, para 85, 24). It is possible to identify this ethos and these values through an examination of the school prospectus. This is a significant document as it is a "crafted, formal representation(s)" of the school (Hesketh and Knight, 1998), so allowing "inferences to be made" about how a school may "wish to be seen in the educational market place." In short, what a school

chooses to say about itself and how it decided to say it—be it informational, promotional, aspirational, or inspirational—in itself tells the prospectus reader something *deeper* about the school. Of course, this impression may or may not be borne out by the reality, and an accurate representation is a matter of integrity. To follow up the documentary analysis of the seven Quaker schools' prospectuses, an interview with a designer of one set of these prospectuses was held. This summarizes what he said:

- Regarding the aim of the prospectuses.

    The aim is to address the kind of parents that might send their children to a Quaker school. People, certainly middle class ones, are used to making choices. It's just a choice you make. However, some of the parents now considering independent education are "first time" buyers and the type of people who'd fly Virgin Atlantic rather than British Airways. The prospectuses are trying to project a particular kind of atmosphere, an independent school with a comprehensive air. So it's been designed to be reassuring but relaxed. They want information. They are, in the main, "Independent" or "Guardian" readers. The Quaker nature of the school is within the prospectuses: they state the Quaker ethos, for example, on the first page, in that the Quaker concern is "that of God in everyone." The prospectuses make a direct reference to Quakerism. So, what is being presented is alternative education: different not intimidating. The overall message is that this school is a quality experience for the child in a way that a parent will want and understand.

- Regarding the format.

    Formats vary, but mainly the format adopted is flexible, easy to update, and of good quality without being too expensive. In one particular instance, the style is not text-intensive but is informative. It adopts a friendly tone, using the second person "you" and the first person "we." In one instance, there are clear and separate sections, using youthful handwriting typescript as section headings. This is supported by the use of lots of photographs of the children engaged in various activities—and looking happy. This supports the message that the school is about people and that a game about status and privilege is not being played.

    Increasingly, the choice of a school is made jointly by the parents and the children themselves, so the focus on the children, mentioned in terms of the photographs, is important. It allows the child as reader of the prospectus to identify with what the school is like or not like. Both of which are important. Though the school isn't in the status game—there are no pictures of impressive buildings and the color scheme used is light rather than the "regal" navy blue or purple—the school badge with a coat of arms effect is present throughout the prospectuses. In one version of the prospectuses the designer had dropped the badge. It has been reinstated as it "gives off" that the school has been there for some time and has stability.

- Regarding the market.

The market is perhaps changing. The school essentially offers a day school with boarding provision. Parents are moving into the area to be near the school as there is market recognition through word of mouth. To overseas readers, the prospectus places the school in a highly attractive country environment and within an English educational system that has a reputation for being excellent. The school is part of this and offers openings into British universities in a caring atmosphere.

Modeling an Alternative Society

The above detail was reported from both interview and documentary sources regarding how individuals involved perceived their schools and how the schools are represented in school publications, with particular emphasis on schools' prospectuses. How close are these perceptions to how things as they *really* are in the schools? Is there a strong and direct expression of the elevated and transcendent values of Quakerism in schools, as the interviewees may claim? Do they really seek to build their pupils positive and active citizenship through links with the immediate environment and wider community? Is this an expression of alternative, progressive, educational values? Or, is mere lip service being paid?

It is not possible to address these questions in detail here. There was certainly evidence from the visits to three of the schools, in terms of the school notice boards and other displays in classrooms, that expressed Quaker values about peace, conflict resolution, and social action. It could be inferred that because of the peace testimony (essentially, attitudes to conflict), behavior that is ignored or tolerated as a "fact of life" in other schools, such as bullying (as sustained repetitive behavior), is an immediate issue to be addressed through action and through adult attitudes and behavior. Of course, anti-bullying strategies are an important part of governmental policies for *all* schools; however, it is asserted here that the persistent incidence of bullying at a Quaker school would *nullify* the reason for its existence. For, what the interviewees would claim (and this seem to be supported by the schools' prospectuses and Web sites and the researcher's own observations) is that Quaker values are openly declared as being the core of their organizational culture.

The first part of this chapter explored how the Quakers as a community are more concerned with culture and actual behavior than theology. The second part of the chapter has focused on that culture as it is expressed within Quaker schools. It has found evidence, albeit limited to the adults running and involved with the schools, that confirms a discernible Quaker school culture.

Again, it is, of course, possible to see such independent schools as an expression of elitism and social privilege. This could be seen as the perpetuation of "middle class, prosperous enclaves" that bus in "nice children" and exclude the many "through cost." In this way, it could be seen as "socially divisive." "Is it all too nice?" These are accusations that Quakers and those working in Quaker schools do not ignore and, as mentioned earlier, are sources of constant debate within the Quaker community. It could be concluded that much of the debate around Quaker schools is not whether they are concerned with a set of elevated values but whether schooling outside the state (public) sector should exist at all.

However, it can be seen that Quaker schools offer significant opportunities for outreach in which a hospitable space is offered. It is here that Quaker values are emphasized and conserved—and so to be brought to the attention of succeeding generations. What is the argument for the continuing existence of these schools? Do they offer other lessons for mainstream forms of education and for alternative educators?

Perhaps Quaker education as an alternative approach can be seen as instilling both hope and resilience. In a philosophical vein, Halpin (2003a) refers to St. Thomas Aquinas (1952) in the identification of the human virtues and bases upon them his hallmarks of hope, that is, mutuality, experimentation, and faith. In a complicated world, there is a need to make children hopeful—in Kantian terms: What can I know? What ought I to do? And for what may I hope? Then, drawing in part from the social sciences, Halpin (2003b) quotes several writers in a variety of disciplines in his discussion of hope. For the social psychologist, Lionel Tiger (1999), hope is what makes tomorrow possible and part of a continuity. For Gabriel Marcel (1951), the social philosopher, hope is only possible at the level of *us* and so offers a release and escape from the selfishness and loneliness of the ego. For Halpin (2003b) himself, schools can be experiments in utopia that project what people *really* want for themselves and society—and offer, again, a positive release or escape from a today that is unsatisfactory and dispiriting, and seemingly inevitable. In short, independent schools that, as was seen earlier, can be placed under the "progressive education umbrella" are an opportunity for values, both individual and societal, other than those of the prevailing hegemony, to be promoted and acted upon. So, while not lightly dismissing the very important and relevant (certainly in England) issues of social segregation and elitism, the positive role of independent schools can be identified. As was seen at the beginning of this chapter in the brief historical overview, George Fox was an anti-establishment figure who did not ally himself and his followers with the existing power structures within seventeenth-century English society. The Quakers, then and now, have distinguished themselves as a group unafraid to "speak truth to power." (The latter phrase is much used about and within Quaker culture. Its origins are usually associated with eighteenth-century Quakers—American Friends Service Committee, 1955).

So if schools, as institutions, can model different values and possible societies, what is offered to the individual child? It can be asserted that with such worldviews come hope, and with that (importantly for the long-term well-being of the child) resilience. So, resilience—the ability to "spring back" from setbacks that a child can experience both within school and in later years—is linked with the way the individual explains themselves to themselves. Further, it links with how explanations are constructed around what happens to that individual in a personal and professional context. An optimistic approach allows that individual to deal with the consequences and emotions of things not going perfectly all the time.

Hope, resilience, and optimism—and a respect for others—are all needed for a future to happen in a positive way. All these attributes are components of action that is aimed at contributing positively to society and helping others. (This is much promoted and documented as being a characteristic of Quaker culture: see Barnard, 1999; Curie, 2007; Gallagher, 2007). For, as the American singer Joan Baez, who has a Quaker background, says: "Action is an antidote to despair." Action as participation and involvement in social and civic matters is also necessary for the vitality and survival of democracy.

Quakers do not believe that they have a monopoly of values and good ideas. Some of what they believe and how they act is likely to be found in other schools and other value systems. However, it can be said that Quaker schools are clearly within the progressive sector of education, one that models alternative societies, different ways of being, believing and living in the hope of friendship with others and the rest of the world—a world in which power is not abused. In that, they demonstrate that the society we live in now is not inevitable and immutable; in short, they allow those sharing their environment to have experience of certain values, to live out another way, and to carry that experience with them for the rest of their lives.

Further Sources of Information

Friends House, 173–177 Euston Road, London NW1 2BJ (www.quaker.org).

The Library of the Religious Society of Friends in Britain and the Quaker Bookshop are both located here. Finding research about current educational matters and Quaker schooling in Great Britain can be difficult. The Quaker Studies Research Association has a database that lists a bibliography that does not include any research into current Quaker educational issues. Its main focus does seem to be historical. Its e-mail address is: www.qsra.org.uk. Twice a year, the Association publishes the journal *Quaker Studies*.

Human Scale Education (www.hse.org.uk).
Quakers in Britain Web site (www.quaker.org.uk).
Quaker Schools Web site (www.quakerschools.co.uk/htm).
Peace Education Network (www.peaceeducation.org.uk).
The Friend, weekly magazine (www.exacteditionjs.com/exact/magazine/385/419).
The Friends Quarterly (www.thefriend.org).
Woodbrooke Quaker Study Centre (www.woodbrooke.org.uk).

Quaker schools in England

Ackworth School, Ackworth, Pontefract, West Yorkshire WF7 7LT. www.ackworthschool.com.
Bootham School, York YO30 7BU. www.bootham.york.sch.uk.
Friends' School, Saffron Walden, Essex CB 11 3EB. www.friends.org.uk.
Leighton Park School, Shinfield Road, Reading RG2 7ED. www.leightonpark.com.
The Mount School, Dalton Terrace, York YO24 4DD. www.mountschoolyork.co.uk.
Sibford School, Sibford Ferris, Banbury, Oxon OX15 5QL. www.sibford.oxon.sch.uk.
Sidcot School, Winscombe, North Somerset, BS25 1PD. www.sidcot.org.uk.

Resources in the United States

With a range of Quaker founded universities, colleges, and schools, much more reading and other resources are available. The Philadelphia Yearly Meeting has a bibliography of American books about Quaker education: see www.pym.org/library/lists/educbook.htm.
Pendle Hill (www.pendlehill.org) is an American Quaker centre for spiritual growth, study, and service. It is situated just outside Philadelphia, PA.

Further Reading

Brinton, H. H. (2007) *Quaker Education in Theory and Practice*. Kessinger Publishing.
Dandelion, P. (2007) *An Introduction to Quakerism*. Cambridge: Cambridge University Press.
Fox, G. (1998) *The Journal*. Harmondsworth: Penguin.
Friends Schools Joint Council (1980) *Quakers and their schools*. London: FSJC.
Hamm, T. D. (2006) *The Quakers in America*. New York: Columbia University Press.

Hay, D. (1997) Spiritual Education and Values. Paper presented at the Fourth Annual Conference on Education, Spirituality and the Whole Child (Roehampton Institute London).
Hubbard, G. (1974) *Quaker by Convincement*. Harmondsworth: Pelican.
Kashatus, W. (1995) The making of William Penn's "Holy Experiment" in Education, *Journal of Research on Christian Education*. Vol. 4. No. 2 Autumn 1995, 157–180.
Quaker Home Service (1986) *Learners All: Quaker Experiences in Education*. London: Quaker Books.
Lacey, P. A. (ed.) *Growing into Goodness: Essays on Quaker Education*. Pendle Hill Publications.
Steven, H. (2005) *No Extraordinary Power: Prayer, Stillness and Activism (Swarthmore Lectures)*. Quaker Books.
Stewart, C. (1953) *Quakers and Education as Seen in Their Schools in England*. London: Epworth.

Chapter 5

A Buddhist Approach to Alternative Schooling: The Dharma School, Brighton, UK

Clive Erricker

Introduction

This chapter will seek to determine the salient features of the approach to education offered by the Dharma School in Brighton, UK.[1] This is not to say that this school represents *the* Buddhist approach to education, by no means. It is a venture envisaged in 1990, which was initiated in 1994, and this is a reflection on its character and achievements in 2007. If you search the web it is difficult to find another Buddhist school that advertises itself. Of course, there are "Buddhist" schools in Buddhist countries, for example, in Thailand, and a Buddhist school in a Tibetan Buddhist oriented subculture, in Dharamsala, India (www.lowertvc.org), but these are not "alternative" schools in the sense evoked in this volume. Besides the Dharma School, the Shambala School in Halifax, Nova Scotia (www.shambalaschool.org), would be the only other candidate that has come to my notice.

Most importantly, we need to ask what can an alternative democratic schooling offer to our mainstream model? What can a Buddhist example offer? These are the crucial questions. Of course, a Buddhist school is a faith school and that, in itself, presents a possible problem. Faith schools are not unanimously renowned for being inclusive, more democratic, or alternative, in a progressive sense. I shall introduce the controversy on faith schools below in order to determine whether a Buddhist school can be positively viewed in the context of this controversy.

Philip Pullman in *The Art of Reading in Colour* begins by stating: "You don't need a belief in God to have a theocracy: Khomeini's Iran is closer to Stalin's Russia than either would like to believe. The real difference between theocracies and democracies is that the former do not know how to read" (Pullman, 2004, 158). The point Pullman is making is that theocracies are to be characterized not by their belief in God but by not knowing "how to read," whereas democracies are defined through knowing how to read. To enlarge on Pullman's point, the important issue is not whether faith, or religious faith, is involved but whether a society is open to or closed to enquiry, open-mindedness, and change. The latter is what makes for a democracy and inclusiveness regardless of religious or nonreligious belief. It is more a question of ideology. The argument that follows is that a democracy is also the only climate in which alternative education can flourish. Therefore, we need to ask of any faith school to what extent it is theocratic (ideologic) or democratic to ascertain its worth as a positive, alternative form of education from which mainstream and alternative schooling might learn.

Arguments Against Faith Schooling

In a recent polemic against faith schools, Roger Marples identifies the main concerns that are the focus of those who are skeptical of those institutions as educational sites within democratic societies (Marples, 2006). He begins, "One of the major concerns shared by those of us with strong reservations about faith schools is that they may not attach sufficient importance to children's autonomy" (Marples, 2006, 22). The autonomy or own agency of young people is a principle concern because, it is feared that, faith schools will wish to prevent or curtail this in accordance with the dogmas that underpin religious belief systems. And, Marples continues, this in turn affects the capacity for independent thought and their potential for growth as individuals: "Those who would frustrate, either intentionally or unwittingly, a child's capacity for independent thought, are denying the child a right to flourish" (Marples, 2006, 23). The matter of intention on the part of the educators, Marples makes clear, is not the main issue, but rather what is imbibed in an unquestioning faith school environment:

> If children do acquire religious beliefs unquestioningly, out of fear or undue respect of parents and teachers, then they may be said to have been indoctrinated whether or not there was any intention . . . and it is unrealistic to suppose that all faith schools would attach priority to ensuring that pupils are encouraged to *critically* reflect on their religious beliefs. (Marples, 2006, 25)

Halstead's requirement that the curriculum should include "education for democratic citizenship" (Marples, 2006, 28; Halstead, 1995) is also mentioned by Marples as an essential criterion for judging faith schools and for doubting their justification.

Richard Dawkins' thinking on this matter chimes with that of Marples. He writes, in the context of addressing the education of children in his chapter entitled "Childhood, abuse and the escape from religion":

> I thank my own parents for taking the view that children should be taught "not so much *what* to think as *how to think*" (Dawkins, 2006, 327),

and that:

> Our society, including the non-religious sector, has accepted the preposterous idea that it is normal and right to indoctrinate tiny children in the religion of their parents and to slap religious labels on them—"Catholic child," "Protestant child," "Jewish child," "Muslim child," et cetera . . . A child is not a Christian child, not a Muslim child, but a child of Christian parents or a child of Muslim parents . . . Small children are too young to decide their views on the origins of the cosmos, of life and of morals. (Dawkins, 2006, 338–339)

It should be noted that Dawkins writes as a scientist and strident atheist, and he is specifically attacking theistic religion. But what is interesting about both Marples' and Dawkins' critiques is the legacy of hostility and assumption that they carry with them. If we wanted to find one quotation that this has emerged from, the following one, from the Enlightenment philosopher Kant, would do pretty well:

> "Our Age is, in especial degree, the age of criticism, and to criticism everything must submit." (Lakeland, 1997, 12)

Critics of faith schools most often start from the legacy that religion is the enemy of criticism and that any form of faith school education will be imbued with this attitude of noncritical nurture. Thus, as an essential vehicle for the Enlightenment project, education must be protected from religious influence. However, this is to take too narrow a view. To be more precise, what critics of faith schooling are really railing against is not faith per se, but doctrine.

Under examination should be the particular concepts of faith, education, and values, and the interrelationship of these concepts that any particular body holds when it decides to set up educational institutions (and this can apply to nonreligious bodies as well as religious ones). Those involved in progressive education have to engage with this issue. In surveying the anti-faith education arguments propagated above, there is emphasis on progressing autonomy, enquiry, and questioning, but not on virtue, compassion, or spirituality: the holistic nature of personhood and our relationship with the world. Faith, in its broadest sense, could be construed as being about just such things. It is instructive to examine the vision behind the Dharma School with this in mind.

The Vision of a Dharma School

The Dharma School opened in Brighton on September 9, 1994 with four children, aged four to six, in the front room of a house owned by one of the parents in Hove, Sussex. In June 1995 it moved to The White House in Patcham, Brighton. Within a year the school had grown to twenty-four pupils and it expanded across the full primary school age range of 4—11 years. It now has a full capacity of eighty pupils between the ages of three and eleven years (Dharma School prospectus). Between a third and a half of the parents of these children identify themselves as Buddhist. The others are drawn to Buddhist values in a broader sense. The school's fees are £4,500 per annum and through these, and with some donations given, it gains its income.

Its inception and purpose were first proposed in a report of a conference on "Buddhism and Education in the UK: A Dhammic Perspective"[2] held at Sharpham House in Devon, UK, during June 7–9, 1991. The subsequent report was called *The Elephant's Footprint: Buddhism and Education in the UK—The Dhamma School Project*. The participants at the conference consisted of four monastics from the Thai Theravadin Forest Retreat Order, who comprised both *bhikkhus* (monks) and Buddhist nuns, plus ten other representatives who were either academics or teachers involved in mainstream or alternative education or were otherwise lay Buddhists. The vision behind the type of education that this conference and its report wished to promote is well expressed in the following extract:

> Two of the seventeen precepts of the Tiep Hien Order founded by the Vietnamese monk, the Venerable Thich Nhat Hanh, are relevant here:
> Do not be idolatrous about or bound to any doctrine, theory or ideology, even Buddhist ones. All systems of thought are guiding means; they are not absolute truth.
> Do not force others, including children, by any means whatsoever, to adopt your views, whether by authority, threat, money, propaganda or even education. However, through compassionate dialogue, help others renounce fanaticism and narrowness. (Carey, 1992, 21)

If we wanted an endorsement of Kant's dictum cited earlier and a response to those who fear the indoctrinatory and uncritical learning that, it is feared, faith schools will promote, it can be found in this quotation from the report.

It is both radically democratic and child centered. It is also very Buddhist, in terms of the original radical teachings of the Buddha that sought to strip away all attachments and regarded all views and opinions as ultimately delusory. The Buddha also taught that compassion and wisdom were the ultimate virtues to be attained. But the Buddha taught over 2,500 years ago and therefore we need to gain a more recent historical perspective as to how those Buddhists who participated in the writing of this report came together. This involves us in identifying part of the way in which Buddhism came to the West.

The Forest Retreat Order was made accessible to Westerners pursuing monasticism by the Thai monk Luang Por (venerable father) Chah. As a young monk he was disaffected with the state of Buddhism in Thailand: the study of doctrine and texts and the laxness of monastic training brought him no closer to realizing the one aim of the Buddha's teaching: the cessation of suffering. His founding of the Forest Retreat Order in the 1940s, following his ordination at the age of twenty-one in 1939, was meant to be a return to the original practice of the Buddha's *Sangha* (monastic community) and its strict rules of conduct, the *Vinaya*. During the course of the 1950s through to the 1970s, Westerners became attracted to the order on discovering it through travels connected with military service (the Korean and Vietnamese wars) and the "hippy trail." This led to a specific monastery being set up for Westerners in Thailand in 1974 and a first Sangha in England in 1977. Both were led by Luang Por Chah's senior Western disciple Ajahn Sumedho.

There is a theme of disenchantment running through this Buddhist history that provides the impetus for a radical renewal of personal vision and educational reform. Ajahn, now Luang Por, Sumedho remarked that "World-weariness and an interest in Eastern religion have a way of breeding good bhikkhus" (Sumedho, 1992, 13). The Buddha's decision to lead the life of a renunciant in India was due to his disenchantment with the householder's life because it could not bring an end to suffering (*dukkha*), the root cause of disease or unsatisfactoriness. Luang Por Chah's decision to set up his own order arose from his disenchantment with the Sangha and state of Buddhism in Thailand. The 1960s and 1970s gave rise to a growing number of young people who became disenchanted with the lack of wisdom and the growth of materialism in Western society, from which some of these "convert" Western bhikkhus, and later Buddhist nuns, emerged.

As different forms of Buddhism have translated to the West, there has also been a vigorous debate on the best form it can take in order to continue to practice the *dhamma* (truth or teaching) effectively. Stephen Batchelor, one of the participants in the Sharpham House conference, has written:

> Just as contemporary agnosticism has tended to lose its confidence and lapse into skepticism, so Buddhism has tended to lose its critical edge and

lapse into religiosity. What each has lost, however, the other may be able to restore. In encountering contemporary culture the dharma may recover its agnostic imperative, while secular agnosticism may recover its soul. (Batchelor, 1998, 18)

Batchelor's disenchantment is with the idea that the practice of the dharma should be regarded as a religious activity or a form of religiosity, with the result that "Dharma practice has become a creed" (Batchelor, 1998, 18) as opposed to "A culture of awakening" (Batchelor, 1998, 113). Thich Nhat Hanh's Tiep Hien Order followed from his activism during the Vietnam War that led to his exile in France, where he founded a rural community based on Buddhist activism with an anti-ideological stance: working on one's own inner development and attending to issues of social and ecological concern.

As a context for the emergence of a dharma school, this history highlights the sort of vision that would inform its purpose.

On page 4 of The Elephant's Footprint it states:

> It is widely recognized that Buddhism has something to offer the West. Its thought is subtle and pragmatic, not at all incompatible with much contemporary thinking; its values and practices, which are designed to develop peace of mind, compassion, care for living things and their environment, are particularly relevant to our time. The question arises as to what provision, if any, we should be making for the younger generation. (Carey, 1992, 4)

The Dhamma School project grew out of the increasingly popular summer camps held at the Forest Retreat Order's main monastery and centre in Hertfordshire, Amaravati (realm of the deathless), which began in 1986. These satisfied the desire for a growing lay following of the Order to bring up their children with an understanding of Buddhist values and started to establish the educational principles on which a Dhamma School would be founded:

> Over the past eighteen months, a variety of ideas and reflections have been articulated touching on key issues, especially the need to give the school a strong spiritual dimension, while avoiding the pitfall of religious dogmatism which usually leads to stultification and separatism. (Carey, 1992, 4)

Key to achieving this was the quality of those in educational leadership roles, who were to "exemplify the teachings rather than merely expound them in a formal fashion [and] . . . emphasis on sila (moral conduct based on the five precepts) both for teachers and pupils, was regarded as crucial" (Carey, 1992 5). As an integral aspect of this process, emphasis is placed on "skilful conduct in everyday life, especially silent observation (meditation), dialogue and enquiry" (Carey, 1992, 5).

The report continues by expressing wider aims in relation to democratic citizenship through:

> ... an education which imparts the skills needed to function in the contemporary world. There is no such thing as that which is not learned. In the West we are also called upon to educate ourselves and our children in ways which enable us to function within society, while at the same time remaining true to our Buddhist principles; not necessarily to conform, but to challenge without becoming alienated from the broad concepts of society. To find a balance (Dhammic "Middle Way") between these two needs is the challenge of the Buddhist educator. (Carey, 1992, 6)

The pedagogical reflectiveness of *The Elephant's Footprint* is strong and focused, mirroring the Buddha's own concern to answer just one key question, how to put an end to suffering. The Buddha's answer, the eightfold path, laid out a means of enquiry toward that end and a spiritual and moral prescription for healthy living. This appears here as a basis for Buddhist education.

> The Noble Eightfold Path ... would appear to be a good and natural model within which to think about Buddhist education ... The emphasis on Buddhist values will be vital in that it will create a properly focused field of action in the world. The exploration of ethics will be the basis for considering how one can most meaningfully gain one's livelihood ... On this foundation of right mindfulness, effort and concentration, the spiritual aspects of human development can be approached. (Carey, 1992, 6)

The document continues by reflecting on what it would mean to become a good learner in this respect, summed up in the Buddhist idea of skilful living:

> A Dhammic perspective of how to become a good learner ... would stress certain values and qualities. The young should be prepared to retain their own centre when faced with shifting values and relationships. The ability to cope with uncertainty, loss, confusion and change, the art of preserving one's equanimity ... is singularly appropriate for the current younger generation ... To learn how to sit with discomfort and risk; to acknowledge dukkha (suffering and general insecurity) these qualities are central to skilful living. (Carey, 1992, 7)

In addition, the report links its vision and pedagogy to children's rights:

> Looking at children's rights in the light of the UN Charter ... we saw how these rights fit into our system under the three headings of provision, protection and

participation . . . (apart from basic provision of food, shelter and medicine) . . . Children need space—internal and external space, uncluttered areas, walls which are allowed to remain bare; space for meditation; a garden or quiet place and time for reflection. (Carey, 1992, 13)

The Research

Fieldwork Methodology

Qualitative interviews were undertaken through visits to the Dharma School on two occasions during 2006–2007. These interviews were conducted with the headteacher and two administrators. Additionally, a group interview was conducted with a year six (10-year-olds) class and two lessons were observed. Subsequently, research was refined by e-mail conversation with the headteacher, Peter Murdock, and Peter Carey of Oxford University, who was instrumental in setting up the Dharma School project.

The Dharma School in Action

You know the most important thing is that the children that come into this school do well . . . if they can't go to the next place and operate successfully . . . then you have failed. With all the goodwill in the world you have failed, you have failed the children. The children have got to thrive. They have got to do well. (interview with present headteacher at the Dharma School, Peter Murdock, March 16, 2006)

Observing the connection between the statement above and the previous vision for the Dharma School is instructive. There is nothing explicitly Buddhist about this statement. Any worthwhile headteacher in mainstream education could have made it. And yet, the point is that such commitment to young people's development has to come from an informed educational purpose owned by that individual. The question that follows is: where did this commitment to educational purpose come from and how will that inform the particular interpretation of what is meant by the generic phrases: "do well," "operate successfully," and "thrive"?

When entering The Dharma School and its environs, especially its classrooms, one is struck by its faith representation. This is eclectic in the Buddhist sense. Each classroom has a shrine with *rupas* (statues) of Buddhas, photos of the Dalai Lama (its patron) and other Buddhist monks and nuns from differing branches of the tradition. St. Francis, from the Christian tradition, also figures prominently. On the walls are examples of

Buddhist teachings and pupil work interpreting verses of Buddhist scriptures as part of their educational development.

However, The Dharma School also follows the English National Curriculum and is inspected by OfSTED (the Office for Standards in Education). Therefore it is committed to the children doing well in their attainment in these conventional academic terms. Nevertheless, from a Buddhist perspective, within which *citta* (mind) does not carry the dualistic assumption prevalent in the West, being separate from heart, there is an understanding that *bhavana* (mental culture or development) only occurs when mind and heart, cognitive and emotional development, are addressed together as a single entity The key to this is the quality of mindfulness. Buddhists are trained to be mindful and that is central to the practice of the Dharma School with its pupils. It is addressed in two interconnected ways. The first is *puja*, a formal practice, which a year six (10-year-old) child described to me as follows:

> A bit like assembly but we sit on the carpet and for one minute close our eyes and stay quiet. Just to calm down and stuff. It is the only minute in the day that we calm down so it is quite special. We enjoy that and sometimes we send good wishes to people like Madeleine McCann[3] in meditation. We discuss openly about issues and if there is a problem Peter [the headteacher] will just talk to us and ask us ideas on how we can stop it.

The child's reference to quiet is significant. To explain further, this relates to the importance of the Buddhist idea of silence; silence is not the brief space we place between speech, for example, like a pause between sentences. Rather, it is the natural space from which speech can be formed. But the value of the speech is dependent on the quality of the silence. It is a reversal of our usual pattern of value. Instead of silence being the necessary pause to take breath (a neutral function), it is the environment out of which positive thought and speech emerge. Silence or stillness is not a gap in speech or movement, rather speech and movement are an intrusion on silence. Thus, silence and stillness are practiced to create positive qualities and release inherent dispositions.

Each puja will have a theme that is designed to ask children to reflect on life and develop awareness—for example, the key Buddhist teaching of life not being satisfactory (dukkha, the first noble truth taught by the Buddha). The meditation, in this example, focuses on something that is not going right, as the headteacher, Peter Murdock, explains:

> So we meditate and I say "What would you say in your life at the moment you are not really happy with?" The children talk about all sorts of things, we did this just last week. One talked about the death of his father, when he

was a baby, another remembered the death of a pet, another said he didn't have a friend, another said I don't like spelling, so you get the extremes. And at the core is dissatisfaction and becoming aware of that. This is consistent with Buddhist teaching in the 1st noble truth: with human life comes dissatisfaction. The Buddha said that we learn from dissatisfaction. We are bringing this fact to the children's attention. In our minds we have dissatisfaction, it is part of human nature, not a problem particular to a child. So they hear other children have the same problem. It is relieving for them to know that the issues in their lives happen in other people's. That is awareness. They are not the only ones that suffer; or their parents aren't the only ones that argue; or they are not the only one who has had someone close to them die. Dissatisfaction is alleviated by knowing it has happened to others and that is why this awareness matters. I would say it is not just a Buddhist teaching, it is a teaching about life for all children who will be able to use it. It is about acceptance of dissatisfaction and furthermore knowing it will change. Change is an important concept in the teaching of the Buddha and, if there is one thing that is constant in this universe, it is change. We talk *about* awareness and mindfulness but we also want to make children aware of themselves. We talk quite a lot about what children are feeling, how they are feeling and sharing our feelings. (interview with present headteacher at the Dharma School, Peter Murdock, June 15, 2007)

Addressing awareness and mindfulness also takes other forms. It also involves acceptance of others as they are and being positive in their thoughts toward them. At the beginning and end of the year the ritual called "sun in the centre" is part of the puja activity. One child sits in the centre of a circle made up of the other children in the class. In turn each child is asked to say something positive to the child in the centre giving him or her feedback for reflection. This is meant to be affirming, focusing on positive attributes. At the end of the year it is based on a memory they hold, positive affirmation in the present and wishes for the future. The child in the middle keeps changing direction to face the child giving feedback. The purpose is:

> For the children to be able to be very direct with one another, be candid but positive.
> It is further teaching method to help children become more aware of themselves, by reflecting on what everyone else has to say to us. Sometimes we think of ourselves very differently to the way others may see us. Some children can be surprised that someone they haven't talked to much in the year thinks very well of them. (interview with present headteacher at the Dharma School, Peter Murdock, June 2007)

The emphasis on positive affirmation is complemented by a strong accent on moral teachings being put into practice through the conduct of everyone in the school. This is based on the five precepts that the Buddha laid down.

They have become the five precepts of the school that the teachers refer back to regularly to remind the children of what is, effectively, the school code. They are:

1. I will not harm another living being.
2. I will not take that which doesn't belong to me.
3. I will not speak negatively.
4. I will try to get on with others and if I do not like another person I will accept that they have an equal right to be in this school.
5. I will take a commitment to try and look after my body and keep a healthy attitude.

In the usual Buddhist way, these are strongly linked to the practice of mindfulness. In this sense they are aspirations as well as principles. They are a code for developing oneself, to encourage reflection, rather than a moralistic stance. The premise for this is that mindfulness affects the moral quality of a person and enables that to be expressed through the precepts.

This is reflected in statements made by the year six (10-year-old) children, such as: "All our teachers tell us not to hurt living creatures. I have quite a lot of friends out of school that think squashing little bugs is lots of fun and I don't find that very good"; and the number of children who remarked on everyone being kind to each other and supportive.

However, this does not mean that the children are saintly. They are regarded as "typical children" and treated as such. The comment often made about small, alternative schools that their pupils will one day have to enter the "real world" is strongly refuted by the headteacher. It is the case that the Dharma School places great emphasis on care, support, attending to the individuality of each child and ensuring their security, and the older children recognize the benefits this has had for them. Nevertheless, it is in attending to how conflict is dealt with, rather than just seeking to avoid it, that its procedures also address.

One example is the strategy called The Box. In every classroom there is a box, into which children daily post any issues, concerns, and comments. These can be positive or negative. The privacy of these comments is respected. The box is a simple cardboard one with no lock and it can be opened easily. Usually at the end of the day it is opened by the teacher or headteacher. Decisions will be made on any issues the comments raise as to whether they need to be addressed or not straightaway. Where it is important to do so, a child will be asked the next day about the comment, and in that way they know it has been read and that they are being consulted on what needs to be done. If there are issues that don't get resolved, for example, between two children, then they are brought together and with rules that have to be followed, such as listening to what the other person is saying, both have to hear each other and try and see the other person's point of view. Both children have to anticipate how the other would prefer the outcome. The aim is to get to the point where they are starting to understand each other. If that can't be done, then they have a stand off where they need, for example, to play with someone else for a while and see how it goes, just like adults. The emphasis is on the process whereby negotiated outcomes can be agreed.

One year-six (10-year-old) child was keen to talk about this procedure:

> We have a box and when we have a problem we write it on a piece of paper and put it in. We can also put positive stuff in, like if we enjoyed a lesson. At the end of the day Peter reads it out to the whole class. But if you put private on it, for example if you are being bullied, he asks you if you want to discuss it with that person. He doesn't read it to the class.

To return to the question of how well the school prepares them for the world they will enter when they leave, in secondary school, and whether they will thrive because they have been well-prepared for it, we can consider the headteacher's criterion that they should be ready to move on. In talking with the children in year 6, typically aged ten years, who will enter secondary school in the following year, they conveyed "mixed emotions." They were excited but also "sad and nervous" and it felt "scary" and "weird." But, in my exchange with them they had obviously become articulate and confident children, for their age, and their openness as to their emotions was evidence of this. Many of the children have spent seven years at the school and it is the only learning environment they have known.

In his assessment of these ambivalent emotions, Peter Murdock speaks of personality types. The ones that are naturally more gregarious in outlook are out of the door moving on to year 7 and 8, the first years of secondary school, and make very quick transitions. They welcome the idea of different things. The deeper, more introverted child makes strong connections and strong friendships. They identify with the past, its images, and relationships and they struggle with that breaking up. If they are more outward in terms of socialization, the transition is easy, and if they are more inward they still transfer with the same strengths and weaknesses but they struggle more at letting go.

For parents the transition can be disruptive as well. As with the children, the aim is to show mutual respect, understanding, and an acknowledgement of individuality to the parents. Consultation with parents is paramount and carried out face to face and through meetings, rather than just sending a letter. The accent is on valuing what parents say and acknowledging any difficulties that new decisions raise. Whenever possible, the decisions will be the result of the consultation itself, thus moving the school forward in an open and transparent way. The school administrator, who has been involved with the school from near the beginning, initially as a parent, records that "We have an openness with the parent community. This is not my experience as a parent of a child now in a secondary school" (interview, June 15, 2007). Another administrator who has been with the school since its inception provided a summary of what it seems the school seeks to achieve overall:

> It is quite a strong community, like a big family, and the children are really happy. It is that sense of family that also can help kids in difficult circumstances. There is a genuine feeling in the children that we really care about them. In my time here I have seen children transformed. They have arrived with difficult behavior problems and within a short time it has all calmed down. (interview, June 15, 2007)

This section has been concerned with how The Dharma School puts provision in place to ensure and evaluate its success. This can be summarized in two ways. In solely academic terms, OfSTED is its arbiter. The OfSTED report of June 12–13, 2007 commented that "It provides a good quality of education" and that "the curriculum is good," citing that the school could improve further in "the use of ICT across the curriculum." In the larger sense of pupils' development, the report underlined the vision of the school being put into practice with the comment that "Provision for spiritual, moral, social and cultural development is outstanding" (OfSTED, 2007). The significance of this report lies in recognizing that, in the larger sense of pupils' development, the processes and mechanisms used by the school to ensure its accountability to parents and children have been endorsed.

An Evaluative Consideration of the Dharma School

A small school originating from a parent base in the front room of a family house in the early 1990s faces many challenges on its way to maturity. In 2007, after thirteen years, an assessment of its success is going to be only partial. In every respect the challenges of these early years depend on aspects of capital. How do you create sufficient economic, educational, and social capital to fulfill the vision? Bea, the administrator, describes the changes as follows:

> I have been here since September 1995. The school moved here from a parent's front room, with originally four children in June 1995. There were twenty children then and now there are 82, including the nursery. We are bursting at the seams. I started as a parent volunteer in the nursery, then as an assistant there. I have seen the school go through lots of changes. It has grown and shifted. I imagine the school like a living person, from toddler through all these different stages. (interview, June 15, 2007)

If we identify some of these changes in more detail, we should arrive at an estimation of the progress the Dharma School has made, its sustainability, and its future potential.

From Charisma to Policy and Procedure

Small visionary schools initially survive through the charisma of their first leaders. Lacking any initial systemic infrastructure and relying on a close-knit body of supporters, they find their way hand to mouth. It was no different in the case of The Dharma School. The present headteacher, Peter Murdock, reflects on those times and the present:

> From Kevin (the previous and first headteacher) to me reflects a difference in the way the school has needed to operate. I have always wanted to move away from charisma toward policy and procedure. You have to have a system beyond personality to flourish. Otherwise, when one person goes the next person is floundering. This is the next step for The Dharma School. When I came I saw the school was too dependent on Kevin and when Kevin left the whole place fell apart. Then they had a terrible year before they hired me. They nearly went out of existence because of this dependency. Kevin had a different era to cover as the school developed and he was flying by the seat of his pants as the school grew. At that time they needed charisma to survive. (interview, July 15, 2007)

The school is moving on from this dependency on charisma, but is not yet fully secure. It also had to move on from the idea of parental ownership. Initially parents had invested in the school, both financially and morally, and expected a return on their own expectations. They were most upset when things didn't go according to plan for them or their children. This leads to conflict in which those who possess more influence can dictate leadership and policy. It is a commonplace occurrence in the development of small, independent schools, but it interferes with the democratic aims of the institution. In the case of The Dharma School it has moved on, and the centre of leadership now resides with the institutional mandate of the headteacher and the governors who are recruited through various Buddhist sources and from parents, and within a clear and agreed development plan. The policies and procedures of the school need to be seen as an implementation of its original vision.

Inclusivity

The school needs to charge fees to survive. The problem is a catch-22. How can you take the school into the broadest social context when parents have to have more than £4,000 disposable income to put their child in the school? It closes the school off to many people. The Dharma School's attempt to resolve this to some degree has been to increase the fees by 15 percent in order to accommodate a bursary system. Present parents who send their children to the school are aware of this. They will be subsidizing other pupils. However, this can only result in a maximum of eight bursaries overall, totaling £32,000. The headteacher's comment on this possibility is, "If we hold it in our hearts to achieve it we could get there."

Non-Buddhist Teachers

The Dharma School's desire to be inclusive extends to the employment of non-Buddhist teachers. It seeks to serve a regional community rather than a Buddhist one. They employ teachers from best-quality applicants rather than on the basis of being Buddhist, though there has to be empathy with the aims and ethos of the school. In this respect there is more work to do in ensuring that all teachers can work toward an ethos policy that can be implemented across all pupils and classes. The basic aim is the teaching of a moral curriculum with a "Buddhistic flavor," for example, in the way that mindfulness is addressed and the *pujas* are conducted, with meditative exercises at their centre. It is still work in progress to determine what it means to have good teachers with clear spiritual affinities with a Buddhist ethos, without a strong Buddhist background, who can carry out the daily teaching of the school. Nevertheless, it is apparent that there are no significant tensions in this respect and that the school has placed itself in a position in which it can address this as part of its progressive development effectively. In the headteacher's, Peter Murdock's, words:

> Those that aren't Buddhist have still got that kind of perception that these are useful tools, a useful way of looking at things, you don't have to believe it, it is not required. With me the four noble truths go pretty deep as part of my faith but for others they can be simply tools to work with and that is fine with me. (interview with present headteacher at the Dharma School, Peter Murdock, March 16, 2006)

Lessons for and from Mainstream and Alternative Education

What Can a Buddhist Example Offer?

I return to the question raised at the beginning of my study. The Dharma School offers a faith-based education focused on progressing the ability of young people in their spiritual development. It does so by emphasizing autonomy, enquiry, and questioning. Its holistic vision of a mature person resides in the idea of moral virtue, expressed through compassionate action in the world toward all creatures and our environment generally. In its educational practice, it has identified an ethos and pedagogical strategies that are based on its vision embedded in everyday routines owned by both teachers and pupils. In this sense one could say it has taken the most significant critical aspects of the European Enlightenment, expounded by the critics of faith schools in the first section of this study, and aligned them with a faith orientation that restores an ethical vision to educational practice based on the Buddha's teaching. In this respect The Dharma School represents a distinctive approach to education and schooling that is often absent in mainstream and other faith-education settings. Critical thinking and virtue ethics have been aligned for a common educational purpose in the development of young people. This contrasts with mainstream schools that pursue solely "academic" excellence with a rule-based normative moral code and faith schools that are catechetically inclined.

How do you create an educational institution that is also a community? This is the central question that The Dharma School has sought to address. It is rooted in the idea that a school needs to be a thriving community in order to benefit the young people it serves. For that to be the case, the school has to situate itself within a vision that the school embodies, which goes beyond the school itself. The Buddhist vision that the school espouses is based on an enquiry into what it means to develop as a person. It is inclusive rather than doctrinaire, and it is democratic rather than insular. It is outward-looking rather than inward-looking in seeking to offer enrichment to the wider society, of which it sees itself as part. It is the thread of these characteristics that gives it both its distinctiveness and its contribution to society as a whole. The lesson for mainstream schools is telling. If you are an institution first and then seek to create a community of the members of the institution without a clear and distinctive vision that will contribute to the wider society, the outcome is likely to be educational failure regardless of the "academic" attainment of your pupils as measured in tests or examinations. The lesson for alternative schools is that if your alternativeness sets your pupils apart from the wider society in such a way as they cannot function within it successfully, you do them a disservice.

Lessons from Mainstream Education

How do you move from charismatic leadership to a sustainable community? Clearly this question has challenged the Dharma School as it has sought to mature. It is invariably the case that the vision lying behind alternative schools can be traced back to a charismatic individual. The problem arises, with institutionalization, that either the vision becomes lost or the dependency on the charisma creates both fragility and, eventually, a form of doctrinal dysfunction because the institution does not have the capacity to change as the society around it changes. From being progressive in its inception, in time, it becomes reactive. While charismatic leadership is often a key inspirational influence, it cannot, of itself, create sustainability. Charismatic leadership is often a sign of the dependency of a community rather than of its maturity. Democratically, this is undesirable. The Dharma School has struggled and almost foundered in seeking to make this transition and, whilst moving forward, still needs to ensure its systems are more robust and its procedures more transparent and accountable in the longer term. At the same time, this might also ensure that The Dharma School's economic sustainability and inclusivity become more assured.

The Question of Faith in Twenty-first-Century Western Education

It is common to speak of those "with faith and no faith." This type of language is damaging. What is meant is those with religious faith and those who are secular. Mainstream schools regard themselves as secular institutions. Faith school is a term reserved for those schools with religious foundations. This reflects a democratically immature mentality. The Dharma School is an example of an institution that exposes that immaturity. It is a legacy of the European Enlightenment that we find it difficult to eradicate. It is significant that a movement born out of a different time and place offers a perspective that is both in tune with the critical approach of the Enlightenment and yet is also firmly rooted in an idea of personhood based on the values of wisdom and compassion, and offers a significant challenge to the way in which we classify society and its educational institutions. It is time to recognize that all young people, whether educated in mainstream-, church-, or alternative schools, need to develop faith. Not faith in something but faith understood as the development of human potential.

Notes

1. Dharma is the Buddhist term for truth or teaching based on the truth, "the way things are."
2. Dhamma is the alternative, Pali language rendition of Dharma, the Sanskrit term.
3. The British girl (aged four at the time of writing) who went missing in May 2007 whilst on holiday with her parents in Portugal.

Further Sources of Information

Bourdieu, Pierre and Passeron, Jean-Claude. (1990) *Reproduction in Education, Society and Culture. Second edition.* London and California, Sage Publications.

Chah, Ajahn. (1982) *Bohdinyana*. Ubon Rajathani, Thailand (available from Amaravati Forest Monastery).
Disciples of Ajahn Chah. (1989) *Seeing the Way*. Great Gaddesdon, Amaravati Publications.
Erricker C and Erricker J. (eds.) (2001) *Meditation in Schools*. London and New York, Continuum.
Erricker C. (2001) "In the Realm of the Deathless: The Thai Buddhist Forest Retreat Order." In *Contemporary Spiritualities: Social and Religious Contexts*. Ed. C. and J. Erricker. London and New York, Continuum.
Erricker C. (2003) *Buddhism*, 3rd ed. London, Hodder Headline.
Erricker C. (2005) "Faith Education of Children in the Context of Adult Migration and Conversion: The Discontinuities of Tradition." In *Spiritual Education: Literary, Empirical and Pedagogical*. Ed. C. Ota and C. Erricker. Approaches, Brighton, Sussex Academic Press, 231–244.
Erricker C. (2007) "Children's Spirituality and Postmodern Faith." *International Journal of Children's Spirituality*, 12 (1), 51–61.
Gyatso, Tenzin (The Fourteenth Dalai Lama). (1996) *Kindness, Clarity and Insight*, New York, Snow Lion Publications.

Web sites

www.dharmaschool The Dharma School, Brighton, UK
www.shambalaschool.org The Shambala School, Nova Scotia, Canada
www.buddhanet.net World wide Buddhist Web site

CHAPTER 6

ISLAMIC SCHOOLS IN NORTH AMERICA
AND THE NETHERLANDS: INHIBITING OR
ENHANCING DEMOCRATIC DISPOSITIONS?

Michael S. Merry and Geert Driessen

Introduction

Muslims have resided in Western countries for decades, yet with the possible exception of Canada and the United States, public opinion, especially that of politicians and employers, for much of the twentieth century was that their residence would be a temporary one. In Europe, labor shortages, decolonization, and family reunification and formation over two or three generations would ensure permanent resettlement. In North America, since the late 1960s, expanded immigration policies have facilitated a large influx of Muslims from the Indian subcontinent and beyond. Consequently, for more than two decades, Islam has been one of the fastest growing religions in the West and is Europe's second largest religion (Merry and Driessen, 2005; Driessen and Merry, 2006).

Though a majority of Muslims continue to use public schools, by the early 1980s a small number of Muslims in most Western countries had begun to mobilize their efforts and establish Islamic schools. The number of Islamic schools in Anglophone countries has been steadily increasing since the mid-1980s, with estimates of well over one hundred in the United Kingdom and more than four hundred across Canada and the United States (Merry, 2007). The majority of Islamic schools in North America and the United Kingdom operate as private or independent institutions, with most receiving no direct state funding or oversight. With this independent status comes tremendous freedom with respect to curriculum, but also for administrative decisions, though schools must comply minimally with a number of safety inspections. Regulatory agencies may either withhold accreditation (United States) or publish criticisms of a school's unsatisfactory performance (United Kingdom), though generally only when schools fail to meet minimum academic standards or other legal requirements are schools effectively closed down. On the European continent, however, there is considerable state funding and oversight, though the amount of funding and the type of oversight varies from country to country. In this chapter we limit our focus to North America and the European Continent, and our European example is the Netherlands. We have chosen the Netherlands for three reasons: (1) there are more Islamic schools relative to other European countries, (2) all Dutch Islamic schools are fully funded by the state, and (3) more data on school performance are available.

Our focus in this chapter is on mainstream, full-time Islamic schools; that is, those that endeavor (even though many fail) to provide a state-recognized education in all subject matter necessary for ordinary primary and secondary level studies. We will concentrate on what they have in common, noting that different schools of interpretation prevail depending on the population they serve, that different cultural and curricular emphases exist between individual schools and administrations, and that a diversity of choices (for example, coeducational or single-sex, primary or secondary) also exist.

Insofar as we assess their strengths and weaknesses, we will show that many Islamic schools minister to the needs of children whose cultural identities are hybrid and fluid, whose religious identities are routinely shunned or derided in Western societies, and whose psychological and physical safety are important for flourishing. Yet we will also consider a number of challenges that many Islamic schools face, including limited resources, inconsistent academic results, and the difficulty of maintaining an Islamic ethos. Next we will also examine whether Islamic schools are inclusive and democratic. Schools that are inclusive will be those that prepare pupils not only to tolerate others but to work toward mutual respect of others who subscribe to a different conception of the good life, and democratic schools will be those that cultivate in pupils the dispositions necessary to construct and sustain liberal democratic values, including, but not limited to, freedom and equality. To illuminate this question we will examine the issue of gender equality and opportunity. Finally, we will consider several best-practices in Islamic schools, many of which they share in common with other denominational schools, and suggest some features that public schools might consider adopting into their repertoire.

Defining Features

Islamic schools in the West aim to provide an environment infused with the teachings and ethos of Islam. In this basic aim, Islamic schools share a similar mission with other religious schools. Though there are many different types of Islamic schools, including varying degrees of orthodoxy, strictness, and ethnic affiliation, many overlapping similarities unite them. To begin with, all Islamic schools aim to promote an awareness of Allah in all that children do and learn. God-consciousness (*taqwa*) is promoted by all of the Muslim staff and is believed to foster the child's development that will lead to inner excellence, justice, and the witnessing to the truth of Islam. Therefore, mindfulness of God is the central aim of an Islamic education, but maintaining equilibrium between the physical and spiritual realms follows from this, and there must be integration and balance of all forms of knowledge. As if to emphasize this God-consciousness, expressions such as *insha'Allah* (if God wills) and *al-Hamdulillah* (thanks be to God) suffuse the speech of school staff (Merry, 2007).

Prayer times in Islamic schools are routine and space is provided for pupils to carry out ablutions either in an adjoining mosque or in the school itself. Friday prayers are typically a community-wide event during which a sermon is given. The language of instruction in most Islamic schools is the official language of the country (and state-sponsored schools require this), though some instruction in Arabic is rudimentary, particularly in religious instruction classes and study of the *Qur'ān*. All prayers are typically recited in Arabic and many expressions, including the universal Muslim greeting, *Salaam Alaikum* (Peace be upon you), are offered in Arabic.

All Islamic schools provide varying amounts of *Qur'ānic* instruction (with recitation), including studies of the life of the Prophet and the period of the first four Caliphs. The moral example of the Prophet Muhammad, whose deeds are collected in the *Sunna*, and whose attributed sayings are collected in the *hadīth*, provides a reliable moral guide. For older pupils, there is also study of jurisprudence, including consideration of Islamic law (*shari'āh*). From these are derived judgments concerning that which is either approved or morally intolerable. All Islamic schools celebrate the two important feasts in the calendar: the Festival of Sacrifice and the Festival of the Breaking of the Fast. Many schools also celebrate a holiday on the birthday of the Prophet.

Besides the usual subjects (for example, maths, science, language arts), art classes are sometimes available, but depictions of persons and animals are strictly forbidden because of Islamic sanctions against idolatry. Islamic songs are permitted, but music classes are available only in a few schools, and many instruments are forbidden. Because Islam compels modesty, dress codes are usually strict not only for pupils but also for staff, and Islamic manners are instilled. Beyond a certain age (typically by age nine or ten), it is characteristic for girls to wear a headscarf (*hijāb*), as a show of inward as well as outward modesty. Gender separation is a common practice in most Islamic schools, at least prior to the onset of puberty. Only in smaller classes, as a practical necessity, does one find the mingling of boys and girls, and even then self-separation tends to happen. Physical education, assuming it is provided, is usually separately provided, according to sex, except in the most liberal Islamic schools, of which there are a few in the Netherlands. In North America, however, most Islamic schools are coeducational.

To the extent that Islamic schools also provide a *cultural* environment that corresponds closely to the ethnic backgrounds of parents, they also share much in common with community- and culturally based education (for example, African-centered, Hasidic schools). Islamic school staff understand their role as not only one to provide academic instruction, but also to provide counseling, role modeling, and spiritual support, both for the pupils and, occasionally, for the parents. Though some changes have occurred in recent years (including a rise in the number of converts), Islamic schools usually attract more parents who are recent immigrants than those who are second or third generation. Particularly in North America, where, irrespective of ethnic background, Muslims are generally well-educated, many Muslim parents freely express their concerns to their child's teacher in Islamic schools, while this is often not the case in the public schools, where recent immigrants—particularly in Europe among the Muslim underclass—do not know how the system works and are unwilling or unable to attend parent–teacher meetings, perhaps for fear of looking incompetent, or because work schedules will not allow for it (Merry, 2005a; Badawi, 2005).

Whatever the case, accountability to the community is normally high, and respect toward adults is expected. Moreover, owing to the stronger formal relations that usually exist between school board members and teaching and administrative staff, there is usually a strong stake in the performance and reputation of the school, as well as the well-being of the pupils. In a number of Islamic schools, school board members are also part of the school management. (This can also create problems, as we will discuss later.) Islamic school teachers, many of whom have considerable teaching experience in other public and private schools, often remark that Islamic school children are much better behaved compared to children in other schools. Staff attribute this to a school philosophy built on inner excellence, a life guided by prayer, morality, and God-consciousness. Issues of faith are broached in the classroom and, theoretically, integrated into the teaching of all subject matter, though, as we previously mentioned, a religiously integrated or

"Islamicized" curriculum remains a formidable challenge for many schools (Merry, 2007). Children and teachers often use their personal experiences as Muslims for instructive examples in classroom discussion. Yet the above description does not apply in the Netherlands, where there is a shortage of Muslim teachers and where the teachers are relatively young and are less experienced (Driessen and Valkenberg, 2000).

Why Choose an Islamic School?

From the perspective of many Muslim parents, public schools are decidedly secular spaces that operate to promote values and teaching contrary to Islamic teaching and practice. Choosing an Islamic education for one's children is therefore a form of protest for many Muslim parents who question the seriousness of multicultural posturing in liberal democracies when the "politics of recognition" (Taylor, 1994) fails to include religious identity. Though some Muslim parents say that they prefer Islamic schools because they plan to return to their countries of origin (Kelly, 1999), the primary motivation for many families is to protect their children from secular influences in the public schools and society, and to cultivate a strong religious identity (Badawi, 2005; Merry, 2005a). For many parents this translates as more *discipline*, particularly when parents may be looking to "correct" delinquent behavior (and incarceration rates among Muslims in Europe are on the rise, see IHFHR, 2005).

When asked what is distinctive about an Islamic education, many teachers and principals report that Muslim pupils feel at peace, that the Islamic school fosters better character, and aligns the actions of pupils with God's will. Many teachers and former Islamic school pupils (some of whom return to teach at their alma mater) say that a feeling of unity and sense of belonging prevails among the student body at an Islamic school. Whether it concerns sharing a dress code, prayer times, eating *halāl* or sanctioned food, or fasting or celebrating Islamic holidays, pupils can feel themselves in solidarity with their peers, and this extends to the *ummāh*, or global Muslim community, beyond the school as well. Being together in the Islamic school means not having to face (at least not as often) bullying, peer pressure, gang culture, harassment, or drugs in a less friendly environment. Other aspects contribute to student well-being, including a higher degree of adult supervision and concern (though as we shall see, this may have inhibiting effects for girls), fewer cliques, and more self-confidence among the student body. Especially from curriculum developers and school principals, one hears that Islamic education offers a structural advantage over Western forms of education owing to its integration in all aspects of living (Merry, 2007).

The core motivations to attend an Islamic school therefore include the formation of a strong Muslim identity and physical and psychological safety, as well as a desire to see improved academic performance. In what follows, we will consider the processes and outcomes of Islamic schools by examining each of these.

Religious Identity

In order to cultivate a robust religious identity, Islamic schools normally operate according to specific religious beliefs and behaviors. Modeling Islamic behaviors is critical both to the ethos of the school and the religious identity of the pupils. Many parents and teachers feel that Islamic values are at odds with the (absence of) morality in the popular culture, but especially in the schools. Islamic schools also strive to provide value coherence between the home, the mosque, and the school, and this is accentuated through the dress code, prayer times, single-sex instruction, and a religiously sanctioned diet.

However, while religious identity is a principal concern, many Islamic schools struggle to provide the requisite religious ethos and role modeling. This is certainly the case in the Netherlands because as many as 80 percent of school staff are non-Muslim (Driessen and Valkenberg, 2000). Though the Muslim applicant pool in North America is usually larger, Islamic schools do hire many non-Muslims who are both certified in the appropriate subject areas and are willing to abide by Islamic school norms. While many teachers also work in Islamic schools because they wish to give back to the Muslim community (and tuition waivers, remitting or reducing school costs for staff members, are an attractive incentive), others do so for smaller class sizes and fewer behavioral problems, or because of the inability to find a position elsewhere, and not because they necessarily wish to. There is also evidence to suggest that schools that receive state funding and oversight have more difficulties fostering an Islamic ethos (Merry and Driessen, 2005; Merry, 2007).

Another factor affecting the Islamic ethos is teacher and principal retention (Uddin, 2004; Zine, 2004). Low retention rates affect the ability of staff to develop and maintain rapport with the pupils, and high turnover rates generally bode poorly for school stability. Retention is a challenge for at least three reasons. First, a majority of Islamic schools in North America are unable to pay their staff competitive salaries relative to public schools; second, many teachers and principals in Islamic schools are acutely overworked. Taken together, this means that only the most committed persons who find intrinsic rewards in what they are doing are likely to stay for very long. Lastly, tensions related to school governance often exist between school staff and mosque leadership concerning the direction the school ought to take. Some schools do operate independently of the mosque but most rely heavily on the support of the local congregation. This may influence the management of the school in ways that impede administrative freedom, though some school staff insist that, whatever the drawbacks, mosque governance is to be preferred to state oversight.

Safety

A larger number of Muslims than ever before have begun to seek out ways to educate their children in an environment in which they will be safe from physical and psychological harm. Psychological safety includes the freedom to openly discuss and explore one's Muslim identity and faith in relation to the curriculum, but also in relation to the wider world. In public schools, Muslim children are often expected to "represent" the entire Muslim community; thus, except where public schools host a critical mass of Muslim pupils, pressures that result from isolation can be immense (Sarroub, 2005; Zine, 2000). Similar to the way that many African Americans suffer "integration fatigue" at one point or another, impelling movement to all-Black communities and schools, Muslims too have begun to rethink the alleged benefits of "integrated" schools (Merry and New, 2008). Islamic schools provide one way for Muslims to show solidarity with other Muslims by freely practicing their religion with peers, and by accommodating or adapting to Western society without being assimilated by its values and expectations.[1]

Harassment, discrimination, and media stereotyping of Muslims have long been commonplace, but following "9/11," threats to the physical safety of ordinary Muslims reached crisis levels. Subsequently, during the autumn of 2001, reports of mob attacks and arson and vandalism on mosques and Islamic schools were fairly widespread. Other acts of terror, including the bombings in Madrid and London, and the murder of the Dutch film provocateur Theo van Gogh in Amsterdam, provoked fear and rage in a vocal minority against the Muslim community. Ethnic profiling, phone tapping, mosque surveillance, and search and seizure style police operations led a number of families to withdraw their children from Islamic schools for fear of reprisals. Adding insult to injury, a "new racism" emanates from a variety of sources concerning discussions about immigration, crime, and job security, including from the Dutch Member of Parliament Geert Wilders and his newly formed Party for Freedom (*Partij voor de Vrijheid*), and neoconservative authors in the United States, such as Daniel Pipes. Charges against Islamic schools in particular have ranged from their being socially divisive to their being Islamic cultural ghettoes that teach hate (Driessen and Merry, 2006).

In spite of these developments, the general trend among (mostly younger) Western Muslims has been toward a religious awakening rather than toward secularism. Many ordinary Muslims who previously had either been nonreligious or complacent about their faith to really discover, or, perhaps for the first time, *rediscover*, Islam, and, importantly, became more eager to inculcate Islamic values and identity in their children.

Academic Performance

While parents certainly hope that Islamic schools will provide an academically rigorous education, most Islamic schools are in their infancy and therefore academic rigor does not yet match school aims. With the exception of a small number of schools in the United Kingdom and North America, most Islamic schools have failed to raise their pupils' level of academic performance (Driessen, 2008; Merry, 2007). Reasons for this range from under-qualified staff and ineffective teaching methods to inadequate facilities. In the Netherlands, all Islamic schools must follow the national curriculum[2] and hire properly credentialed teachers. Even so, only slight gains, as compared with other schools, have been documented. We will provide more detail on achievement a bit later. For now, we will briefly highlight several compelling reasons to help explain the unimpressive academic performance of most Islamic schools. Of course, "unimpressive" must be understood against the high ambitions and expectations Islamic school boards and staff set for themselves.

First, a strong correlation exists between the socioeconomic status of pupils and their school achievement. Without exception, all Dutch Islamic schools are dominated by large concentrations of pupils from disadvantaged backgrounds, and this means that the social and cultural capital that many other, mainly middle-class, pupils use to their advantage is comparatively diminished among Muslim children. Related to this is the lower education level of most Muslim parents (as this bears upon the ability to assist with schoolwork, advocate for their children at school, etc.), the generally low proficiency in the language spoken at the school, which by law must be Dutch, and the fact that very few Muslim parents in the Netherlands who select Islamic schools are believed to be integrated into mainstream society. Further, while many Dutch Islamic schools have promised higher parental involvement, only modest gains have been observed (MinOCW, 2007a; Driessen and Valkenburg, 2000).

We may also mention teachers' low expectations of pupils' potential. Muslim and non-Muslim teachers alike in Islamic schools lack the same level of experience as teachers in the public schools. Particularly, young and inexperienced teachers are vulnerable both to the cynicism of their older peers and to the challenges they face in large classrooms with higher concentrations of children from disadvantaged backgrounds (Merry, 2005c). These lowered expectations are prevalent in many urban environments with high concentrations of socioeconomically disadvantaged pupils. But other reasons include (1) the disorienting effect of having a majority of non-Muslim staff (as in the Netherlands) in a somewhat artificial Islamic school environment; (2) conflicts that frequently arise between the teaching staff (most of whom are secular or Christians) and the more orthodox school board; finally, (3) both the mismanagement of some Islamic school boards and the rather hostile political and media reaction to Islamic schools. All of these help to explain why, after twenty years and relative to average Dutch schools, only very modest success rates can be reported of many Dutch Islamic schools (Driessen, 2007).

Conversely, in the United States and Canada, both socioeconomic and education levels are generally higher among Muslims, meaning that except in the more remote areas there is less difficulty in recruiting Muslim teachers and administrators with a university education, or with a state-issued license. However, because most North American Islamic schools do not receive direct state funding and little oversight exists, academic offerings and pupil performance are inconsistent and varied. Though they are few in number, the most successful Islamic schools operate with administrative autonomy, a balanced budget, excellent school facilities, highly qualified staff, and competitive academic outcomes. Yet despite the higher socioeconomic status of North American Muslims, most Muslim families are unable to afford the tuition fees Islamic schools charge, without tuition reductions, tax credit options, or scholarships.[3] Many families do make enormous sacrifices in order to send their children to private schools, and in some places (for example, Milwaukee) vouchers may be used to pay for all or at least part of the tuition of participating private schools, including some Islamic schools (Merry and Driessen, 2005).

Achievement in Dutch or North American Islamic schools is not tied exclusively to funding or availability of resources. Yet because many Islamic schools are forced to hire under-qualified staff and operate with limited space and curricular resources, academic achievement is normally affected. Further, most Islamic schools experience some tension between the desires of the parents, the school board, the administration, and the teachers. All of these undoubtedly hinder effective school performance. Particularly among the upwardly mobile, Muslim parents more frequently place their children in schools that will ensure their success than gamble on the offerings of the local Islamic school. Consequently, many Islamic schools must also endure the skepticism of the Muslim community, most of whom continue to question the present ability of new Islamic schools to deliver on claims of academic excellence. In the next section, we will return to the question of academic performance and examine specific outcomes more carefully.

Challenges Facing Islamic Schools

Resources

While the number of Islamic schools continues to climb, the greatest challenge facing Islamic schools in North America is limited resources. Presently no Islamic schools in the United States, and only a handful of schools in Alberta and Quebec, Canada, are eligible for direct state funding (Trichur, 2003). Whether it is inadequate facilities and space, poor provisions for health, safety and general welfare of pupils, or an operating budget that precludes hiring the most qualified persons to teach (and a large percentage of teachers and school administrators are sorely underqualified), many Islamic schools face a tenuous future. In fact, a significant number of Islamic schools close their doors within a few years, and a small number have been forcibly closed by the state, as recently happened in Amsterdam where three schools were closed in 2007 for embezzlement and mismanagement. The image of Islamic schools has worsened in 2008, following reports by the Inspectorate of Education, which revealed that at least one Islamic school in Rotterdam (Ibn Ghaldoun) used state monies to pay for pilgrimages to Mecca and Medina not only for the school staff and pupils but also for parents and complete outsiders (MinOCW, 2007b; Elsevier, 2008). Yet notwithstanding these recent scandals, in the Netherlands, where funding and oversight are provided to all schools that comply with state requirements, Islamic schools generally have enjoyed stability, though strict guidelines concerning curriculum and teacher qualification certainly curtail school autonomy. However, without state funding, Islamic schools in the Netherlands would probably not exist, as the majority of Muslims are without the means to pay school fees.

Schools Boards, Principals, and Teachers

In the Netherlands, the most formidable challenge continues to be the composition and orientation of the school boards. Many school boards are not familiar with or sensitive to legislative requirements, have had no training as school managers, have inadequate educational background and low levels of proficiency in the Dutch language, and are principally oriented toward their own ethnic minority and not toward the broader Dutch society (Dumasy,

2008; Driessen and Merry, 2006). Finally, some members of Islamic school boards assume the position because of its status and power and, sometimes illegal, financial gains.

In North America, many school teachers and administrators have no formal training in education, and a severe shortage of qualified teachers persists. Many principals do have previous managerial experience but have little experience or training in education. Further difficulties may arise between the often conservative or more orthodox school board and the (sometimes non-Muslim) school staff concerning the most desirable processes and outcomes of the school. All of this means more difficulties for schools in terms of recognition, whether by the state, accrediting agencies, or simply skeptical parents. Given the poor quality of many Islamic schools, a select few educated Muslims continue to vocalize strong opposition to Islamic schools as a disservice to Muslim children (Kabdan, 1992).

As we previously noted, many teachers do have previous teaching experience, which gives them a vantage point from which to compare their classroom experiences, though some have only the worst experiences to draw upon (for example, behavioral misconduct). These negative impressions sometimes are passed along to their pupils, that is, that public schools are unsafe and academically undemanding, or that public school pupils are sexually promiscuous and motivated only by materialistic pursuits. Further, though Islamic schools often use similar textbooks to their public (or private) school counterparts, these very textbooks are believed by some to infect Muslim youth with doubt and moral corruption. Because very few textbooks exist that have an Islamic perspective (though this is now changing), teachers are encouraged to critically examine existing curricula, syllabi, and textbooks in order to make the revisions necessary for reflecting an Islamic view of humanity as taught in the *Qur'ān* and the *Sunna*.

Parents

As we have seen, parents select Islamic schools for a variety of reasons, including religious identity, cultural coherence, academic rigor, and safety. Some believe that to send one's child to a public school is akin to sending "lambs to the slaughter," as public schools are thought by some to insidiously promote "secular humanism." Therefore, Islamic school selection for many religiously conservative Muslim parents is motivated by a need to shelter and discipline their children from developing an interest, or participation, in clothing fashions, makeup, tattoos, material possessions, and relaxed attitudes toward one's elders, which are seen as evidence of vanity and amoralism. Indeed, the Western practice of romantic dating alone is enough for some parents to send their children to an Islamic school, even when they may offer up other reasons for their selection. Sex education, art education, and exposure to popular music and dance in public schools are other reasons.

Among many parents there is the strong preference for single-sex schools. Accordingly, Islamic schools, for some, are a sanctuary that either will protect their children from harmful secular influences and contact with the opposite sex, or simply help their children to "shape up" and learn the morality and customs of their parents' home culture. Yet more than cultural continuity, for more conservative parents, Islamic education is the principal means of combating godlessness in the world. For critics, which includes some Islamic school teachers, these sheltering tendencies raise questions concerning whether children will be prepared to transition to other schools "outside the bubble," including whether they will possess the democratic dispositions necessary to respectfully engage with the wider culture.

Parents' involvement in their children's education at an Islamic school inclines toward the extremes. In North America, parental involvement among the upwardly mobile, owing to extremely high expectations and educational background, may border on the intrusive, whether the meddling concerns higher academic performance or a protest against that which is seen to be "un-Islamic." Conversely, in the Netherlands, different orientations toward the school, lower education levels, or geographic distance keep many parents from being more involved. A lack of proficiency in Dutch on the part of the parents also poses another barrier, especially as a majority of teachers are native Dutch.

Achievement

Reliable school achievement reports from North America are few and far between, and much of the data are self-reported by Muslims invested in the success of Islamic schools (Mohamed, 2005). A number of well-established schools with large enrollments self-report their student test achievements online and draw attention to their graduation rates and percentage of university-bound students. However, as most Islamic schools in North America are private and receive no direct state funding, beyond the requirements of the accrediting agencies and health and safety inspectors, they are not obligated to report school performance. Consequently, self-reporting continues to be a highly variable and anecdotal phenomenon. Further, a majority of Islamic schools are simply struggling to get by and, for the time being, are unable to produce the results they aim to (Merry, 2007). Conversely, in the Netherlands a number of quantitative analyses have been conducted comparing the performance of Islamic schools to reference schools (schools with a similar demographic) as well as to other state-funded public and denominational schools (Driessen and Bezemer, 1999).

After analyzing data from the 2002–2003 and 2004–2005 academic years and comparing these with earlier analyses from 1994–1995, 1996–1997, and 1998–1999, Driessen (2007) concluded that with regard to language, maths, and reading skills, on average Islamic schools perform at least as well as reference schools, that is, schools with a comparable socioeconomic and ethnic student population. In some instances Islamic schools perform a little, but not significantly, better than reference schools. However, their achievements are far below those of pupils at other, average, Dutch schools. These findings apply equally to all of the five school-years studied, which means that the Islamic schools' goal to raise the academic level has not been realized in the past ten years. With regard to noncognitive factors (well-being, self-confidence, social position, etc.), there are no discernable differences between Muslim pupils at Islamic schools and those in other schools. Also, the fact that the well-being and self-confidence of pupils at Islamic schools does not differ from that of pupils at other schools does not support the idea that once Muslim children are in a safe or culturally coherent environment of their "own" schools, their well-being or academic performance will improve.

However, there is no question that many Islamic schools are performing academically well. Among well-established North American Islamic schools, self-reporting on school Web sites of high academic scores on standardized tests, the Scholastic Aptitude Test (SAT), and college attendance rates among graduates is common. In these successful schools, a majority of pupils come from well-educated—though not necessarily affluent—families.[4] Well-resourced schools that have managed to procure adequate facilities, a qualified and devoted staff, and pupils from more middle-class backgrounds generally perform at least as well if not better than other private religious schools.

Islamic Ethos

The idea of an Islamic ethos continues to puzzle many Muslims, including those who work in Islamic schools. In North America, while the Council for the Islamic Society of North America (CISNA) provides guidance for establishing an Islamic school, and foundations such as IQRA book publications supply teachers and principals with a variety of learning materials, no central agency exists through which Islamic schools might train teachers or organize their instruction. This means that Islamic schools frequently operate in isolation from each other, subscribe to different administrative styles, and are subject to the regulatory requirements of different accrediting agencies.

In the Netherlands, Islamic schools operate under an umbrella organization known as the Islamic School Board Organization (ISBO). Additionally, an Islamic Pedagogical Center in the Netherlands develops Islamically appropriate teaching and learning methods. Following the negative publicity that some Islamic schools received in 2007, the ISBO asked for and subsequently received more authority to intervene in cases of school board mismanagement.

The fact that many teachers in Islamic schools are themselves not Muslim and the textbooks that most Islamic schools use are not written from an Islamic point of view also raises questions about the extent to which Islamic schools can cultivate a strong religious identity. In an attempt to maintain strong role modeling, some Islamic schools enforce behavioral contracts with their teachers so that infractions outside of school (for example, smoking, consuming alcohol) will result in employment termination. Even so, many parents who select Islamic schools do not themselves practice Islam, or know very little about their faith. All of this suggests difficulties for the cultivation of an Islamic identity of some Muslim children, who cannot help but notice the inconsistency (or hypocrisy) of their parents.

Yet other vexing questions arise vis-à-vis an Islamic ethos. Muslim educational philosophers argue that an Islamic education unifies all striving, and all learning, as a spiritual endeavor (Merry, 2007). In other words, there can be no secular pursuits, for all of life must be integrated through faith. Nevertheless, "Islamic studies" beyond readings in the *Qur'ān*, *hadīth*, or *Sunna* remain a vague field of inquiry in a majority of schools and, in most schools, operates as a separate course of study distinct from maths, science, and literature (Badawi, 2005). Consequently, many teachers wonder whether an Islamically integrated curriculum denotes anything meaningful, or, whether it is even possible. Partly, this is because no training center exists that supplies teachers with the knowledge base or pedagogical skills they need. Some teachers use imported materials that address different cultural norms and expectations; however, a great deal of confusion and debate exists over what is acceptable (*halāl*) with what is unacceptable (*harām*) behavior and practice. This is particularly contested territory because cultural practices and beliefs often intermingle with religious practices and beliefs, and separating the two is a messy affair. Further, because literal (though not necessarily political) interpretations of the *Qur'ān* are commonplace, and because many issues (for example, stem cell research, the use of violence against non-Muslims) on which there has been no religious ruling (*fatwā*), now confront Muslims, teachers often find themselves unable to answer the questions their pupils put to them. Young people often turn to Islamic chat rooms or seek out their own answers from teachers or imams that they respect rather than listen to the advice of an "imported imam," who is typically trained abroad and cannot relate to Western youth. While their numbers are growing, the shortage of Western-born and raised imams and teachers remains acute.

Are Islamic Schools Democratic?

The fact that religious faith and democratic engagement with one's host country can operate in tandem should register little surprise. Many social policies and political arrangements have been advanced by persons espousing strong religious beliefs. Though religion and democratic ideals do occasionally come into conflict, especially as it concerns exclusive and absolute truth claims, religiously inspired ideals need not be at loggerheads with the virtues of liberal pluralism, including mutual respect and basic human freedoms (see De Ruyter and Merry, in press; Merry, 2004). The same holds true for education, to wit: though some methods of religious instruction undoubtedly impede rational and critical thought, one should not assume that dogma occludes rational thinking, or that fidelity to religious tenets precludes the consideration of dissenting views. Religious persons can be perfectly reasonable citizens, capable of engaging with others whose views differ manifestly from their own. So how does this bear upon Islamic schools?

As we previously noted, many Muslim parents select Islamic schools in order to minimize the exposure of their children to secularism or views believed to be hostile to Islam. Yet these motivations are curiously juxtaposed with other, more material, pursuits. Thus, in her study of Sunni Muslims in Chicago, Garbi Schmidt found that many Islamic schools use the rejection of American society to legitimize their existence, while in practice

> they are forced to include aspects of American society, because the curriculum must satisfy parents' academic ambitions for their children as much as parental desires for an "Islamic" environment. [Islamic] schools, therefore, become American institutions. (Schmidt, 2004, 81)

Islamic schools are therefore sites for contestation and dispute, for they strive to address a number of, sometimes conflicting, aims. For example, Islamic schools aim to provide academic rigor so that pupils will have the skills and qualifications to succeed both in secondary and tertiary education as well as the labor market. Yet Islamic schools also aim to cultivate robust religious identities that will enable pupils to challenge the very trappings of material success that many of them will inevitably enjoy.

All of this is perfectly consistent with the stated aim of most Islamic schools to promote dual citizenship: one to the global Muslim community (*ummāh*) and one to the wider society from which its pupils are drawn.[5] Thus, in theory, *da'wa* or witnessing to one's faith is consonant with teaching civic virtues such as respect toward others. Of course, respect for those who espouse a different reading of Islam, or who are non-Muslim, will be more evident among those with whom there has been positive interaction. Yet for the moment it remains unclear just how much interaction pupils in Islamic schools have with "otherness." This is so for at least three reasons: (1) Some Muslim parents who select Islamic schools are often reluctant to allow their children to form close friendships with non-Muslims or even with Muslims from different cultural, racial, or theological backgrounds; (2) analogously, "protection" from different points of view, or, if one prefers, cultural coherence, is partly the *raison d'être* of Islamic schools; indeed, Islamic schools exist at least partly in order to *counter* the prevailing cultural attitudes in liberal democratic societies; (3) finally, the enrollment of non-Muslims in Islamic schools remains very low; indeed, the overwhelming majority of Islamic schools contain no non-Muslim pupils. All of this suggests far less contact with difference for a majority of pupils in Islamic schools, especially in the early years when paternalistic control over what children do is more stringent, and limited contact portends worrying trends for inclusive and democratic education. Be that as it may, being situated in a liberal democratic milieu has its own effects. One cannot absolutely eclipse all interactions with the outside world, try as some might to do so.

Importantly, Muslim pupils raised in Western societies, particularly by the time they reach adolescence, expect that reasons or justifications ought to be given, and that blanket authority is insufficient and even unacceptable. As Mazen Istanbouli found in his study of an American Islamic high school, many pupils are not afraid to question the principles of Islam or the authority of their teachers. He quotes a school administrator, who says,

> If you question certain principles of Islam in certain communities, they call you a heretic and they will attack you sharply. I like the fact that [our] kids are a lot freer and they question everything. A lot of that is found in the early Islam but not in the Muslim communities now spread all over the world . . . [our kids feel] that "so what if you are the teacher, unless you earn [our] respect, I am going to question you." And "so what if you are the principal." I feel that this is the right approach. When you deal with those kinds of kids, you respect them and you earn their respect. You produce leaders and not sheep. (Istanbouli, 2000, 220–221)

While some studies (Oriaro, 2006) have suggested that democracy is foreign to the idea of an Islamic education, many Islamic schools aim to promote these and other democratic qualities, such as exposure to different points of view without deriding those views as irredeemably wrong. This exposure may come in the form of guest speakers (including rabbis and priests) or class visits to different school environments. The more this occurs the more one can expect Islamic schools to enhance, and not inhibit, democratic dispositions. As the foregoing quote suggests, many pupils of Islamic high schools, precisely *because* they are raised in a Western context, are generally freer to question and challenge traditional thinking when no reasons are given, and to rethink how to be a Muslim from a different cultural or political point of view. In many respects, pupils in Islamic schools—again, in the higher grades—can be expected to reflect upon their beliefs to a higher degree than children who accept mainstream values, for their commitments will evidently be at odds with much of what the larger society values and this will require greater attention to the *reasons* for maintaining those differences. In a small number of Islamic schools, Muslim pupils are even given the freedom to openly question the dictates of faith. These trends are consistent with the democratic dispositions liberal democracies prize (Merry, 2007).

Yet while many try to carve out a place for critical thinking and reinterpretation without charges of heresy or "innovation" (*bid'a*), it is difficult to foster critical thinking when a critical mass of teachers recruited to teach in Islamic schools adhere to a more authoritarian, teacher-centered pedagogy or when chauvinistic practices that pass for "tradition" undermine democratic ideals such as mutual recognition and the freedom to dissent. (The same is obviously true in many non-Islamic schools.) To focus this discussion further, we will briefly consider gender-differentiated treatment in many Islamic schools. Gender equality matters if democratic forms of schooling are to prosper, because it remains inescapably true that education that places severe restrictions on one's freedom of conscience, movement, expression, and association, even when these are motivated by a well-intentioned theology of complementarity, that is, "equal but different," diminishes the possibilities of promoting democratic dispositions in many Muslim pupils.

Gender Equality

The question of gender equality will either elicit yawns or irritation from many Muslims, who think that Westerners are obsessed with gender, or that they hypocritically point the finger at Muslims without acknowledging the differentiated treatment of women in other religions or in Western society. We are under no delusion concerning equality of the sexes in Western culture. Notwithstanding the tremendous gains feminists have made over the past one hundred years, legislation to actively promote equal rights for women in Western countries is indeed a fairly recent phenomenon and the "glass ceiling" remains implacably in place in many domains, including in some public schools.[6] Muslims are correct to point both to this historical record (including the right of women to vote, which came later in Western countries than in some Islamic ones) and to the questionable belief that sexual freedom is tantamount to equality with men. But the failure of liberal democracies to consistently live up to the democratic ideal of equality of opportunity for women is no argument for perpetuating the practice of gender inequality under the guise of religiously circumscribed gender roles. For while Muslims believe that Islam breaks down many national and ethnic differences, differentiated gender treatment continues to be a troublesome issue in many Islamic schools, notably for many of the female pupils. Yet a critique of gendered practices need not originate outside of the Islamic school; there is considerable anecdotal evidence to support the claim that challenges are emanating from *within* the Western *ummāh* (see Istanbouli, 2000; Manji, 2003; Zine, 2004).

A critique of gendered practices does not have coeducational classrooms as its aim. A compelling literature exists pointing to gains for many girls who are educated apart from boys, for reasons having to do with better self esteem, higher participation levels, and generally more emphasis on learning versus physical appearance (Stabiner, 2000; Halstead, 1993). Single-sex schooling has a long history in Western liberal democracies (notably in the United Kingdom), and many school districts in the United States are also beginning to experiment with single-sex classrooms at the middle school level. All of this suggests not only that the question of single-sex learning per se is a red herring, but that a sensible rationale may be given for separating boys from girls during adolescence. Yet critics of single-sex education in Islamic schools, including many staff and pupils, sometimes ask, "how are boys and girls supposed to learn to interact with each other when they are kept apart for much of their childhood and adolescence?" This is a practical concern, one that challenges the idea of separateness as liberating when it often leads to inhibition around those of the opposite sex. However, more worrying is a corresponding differentiated gender *treatment*, particularly when boys are granted far greater freedoms than girls, including, in many schools, opportunities to engage in athletics or socialize outside of the school.

Despite claims of equality (cf. *Qur'ān* 33:35), Muslim girls are usually more restricted than boys in the freedoms that they enjoy—inside and outside of school—and in the expectations that many feel imposed on them to become mothers or to forgo a career. Limited opportunities are more common among girls who have parents with little education or who follow the customs of another culture that constrain what girls can do; the pressure to marry young and to begin conceiving children continues to be a strong expectation for (Hermans, 1995; Haw, 1994). Careers, for example, usually must take a subordinate position to family duties (including for many the obligation to care for the husband's family as well), and leadership roles, especially in Islamic affairs, are normally assumed to be the domain of men.

So the practice in nearly all Islamic schools of limiting contact between the sexes, of assigning leadership roles to boys, and of granting general freedoms to boys while denying them to girls, appears—certainly to more than a few Muslim girls—to limit their inclusion and democratic participation. Perhaps nowhere is their disempowerment more readily felt than in the way Muslim females are "protected" by men. This happens mainly because Muslim females have come to visibly embody the very essence of Islamic piety. Muslim females are expected to exert ceaseless vigilance over their modest appearance lest they be blamed for usurping male authority or for unwittingly enticing men.

Therefore, gender-differentiated treatment for female pupils is something of a paradox, for while girls often feel more relaxed or supported in an Islamic school due to the fact that they do not stand out or look "odd," they also face greater restrictions on what they can do (Kelly, 1999; Zine, 2004). Limitations on what Muslim girls are permitted to say or do have important implications not only for the breadth of experiences and aspirations that female pupils are permitted to have, both inside *and* outside of school, but also for the range of democratic freedoms Islamic schools are purported to uphold.

Conclusions

There continues to be a lot of variation in Islamic education as it presently exists in the West. Differences exist between countries, of course, but also within countries. Socioeconomic and educational backgrounds between Muslim populations in Europe and North America are noticeably different. Even poorer Muslim parents in North America generally are considerably more educated than their European counterparts (Badawi, 2005). Islamic schools also vary somewhat in their administrative and pedagogical practices, and even the religious orientation of the schools is more orthodox or strict with some than with others, including whether girls enjoy the same opportunities and freedoms as boys.

However, all Islamic schools enlarge upon the purposes of education by seeking to form strong religious and cultural identities, to raise academic achievement, to promote higher parental involvement and community support, and supply pupils with a safe school environment, both physically and psychologically. Many schools manage to produce impressive academic outcomes, civic engagement, multicultural awareness, and above all, a strong moral foundation. Many Islamic schools also enhance democratic forms of education inasmuch as they provide a space for Muslim children to freely question what it means to be Muslim in a society that is frequently hostile to Islam. Liberal democratic virtues have less to do with *which* view one takes and more to do with the *manner* in which one holds and expresses one's views, that is, in mutually respectful tones with others who adhere to different points of view. Such is the value of liberal pluralism. Inasmuch as Islamic schools foster these outcomes, other schools ought to take note.

Perhaps especially, other schools ought to take note of the fact that the Islamic school alternative is appealing precisely because public schools—in the minds of many Muslim parents—fail to provide these benefits for their children. Of course, particularly in North America, public schools have many more demands placed upon them than private schools—they are, after all, answerable to the public and not simply a paying clientele—and hence what some parents expect public schools to do is often unrealistic. For example, many secondary schools are simply too large to provide consistent role modeling or sense of community that smaller schools do. Individual teachers often show the care and concern of their private school counterparts, and high teacher expectations are not uncommon. However, public schools must contend with bureaucratic institutional structures, testing regimes, teacher unions, public accountability, and so on. Consequently, Muslims and other religious minorities must be realistic about their demands and recognize that religious recognition in public schools will not always happen in ways that measure up to what religious school environments can provide (see Lee and Smith, 2001; Toch, 2003).

Even so, one of the most important lessons to take away from this chapter is that public schools must remain open to criticism and dialogue. Muslims ought to be free to make reasonable requests for accommodation, and public school administrators and teachers need to work harder to better understand the Islamic faith and the cultures Muslim children come from in order to minimize the cultural alienation many pupils purportedly feel while in school. Criticisms of multicultural education and even multicultural policy are fair when they expose the fact that religious identities are being deleted from the politics of cultural recognition (Trichur, 2003; De Ruyter and Merry, in press). For many cultural minorities, religion is integral to their core identities; indeed, it influences all that they think and do.

Aside from concerns about gender equality, criticisms of Western Islamic schools have principally focused on concerns about ethnic and religious self-segregation, an inordinate emphasis in some schools on Islamic studies and *Qur'ānic* memorization, and, in some schools, an inadequate preparation for life in a multicultural society. So long as a large number of Western Muslims continue to feel alienated from the societies they inhabit (exacerbated by high unemployment, poverty, and ethnic profiling), the challenges of integration—particularly *psychological* integration that belies a sense of belonging—remain. While radicalist sects like *Hizb ut-Tahrir* have flourished at the margins of Western society (Husain, 2007), their presence is *not* strongly felt inside of Islamic schools. Importantly, radicalization is, we contend, a more likely phenomenon *outside* of Islamic schools because the environment of confrontation and alienation that drives young men into extremist movements is largely absent in Western Islamic schools.

In North America, for the moment there is very little public concern vis-à-vis Islamic schools, in part because most operate very much like other private religious schools (that is, receiving little or no direct state aid), but also because the Muslim population is more educated and appears to have integrated itself into the mainstream culture to a significant degree. However, there are reasons to be concerned about the quality of education at many new Islamic schools, especially when facilities remain inadequate, many teachers remain underqualified, and even the most well-funded and well-staffed schools struggle to continually improve and retain qualified teachers and administration. Finally, our discussion in this chapter concerning gender equality has served to illustrate the distance that most Islamic schools have to go to provide an educational space in which *all* pupils are invited to cultivate democratic dispositions.

In the Netherlands, because all schools, both public and private, receive equal funding, financial worries are not the primary concern. Rather, in the wake of the "9/11" attacks and the murder of Theo van Gogh, and more recently the scandals involving corruption and embezzlement, public hostility toward a vocal or visible Islam in the Netherlands poses a mounting challenge to the existence of Islamic schools. An increasing number of politicians openly advocate a ban on Islamic schools. They are being supported in their crusade by a ceaseless stream of negative media attention. Some of this attention admittedly is deserved, as a number of schools were forced to close their doors because of poor enrollment, low academic performance, or fraudulent and failing school boards (Driessen, 2008; Dumasy, 2008). In light of these developments the attention for the moment seems to be aimed at internal frictions and at averting one crisis after another. Consequently, not enough time and energy remains for improving the education process itself. Given the present difficulties facing a growing number of schools, the reputation of the Islamic school sector as a whole apperas to be under threat.

Yet even larger challenges remain. With increasing instances of "homegrown" terrorism on its own soil, Western governments are, for the moment, becoming less and less tolerant of a visible Muslim minority, and many legal obstacles will be increasingly placed in the path of those who wish to establish or maintain Islamic schools. Particularly in the Netherlands, Muslims have been expected to prove their "democratic competency" or "civic preparedness" to a degree not expected of citizens of other religions. (Indeed, Hindu schools in the Netherlands, and more recently a voluntary-aided Hindu school in west London, have been well received.) We have noted a variety of ways in which Islamic schools enlarge upon the purposes of education, yet whether all Islamic schools will be sites for democratic learning where pluralism is valued remains to be seen. Some schools are like this, but others are not. To reach these goals—which, it should be noted, Islamic schools normally set for themselves—Islamic schools, too, must remain open to criticism and dialogue. If Islamic schools fail to provide academic excellence (and some do fail), or merely exchange one type of conformity (peer pressure in public schools) with another of their own (unquestioning acceptance of rules or gendered expectations), there is reason to question whether they are meeting their own standards of excellence and equality, let alone those championed by liberal democrats. Only time will tell whether most Islamic schools can live up to this challenge. The conversation has already begun and will continue, but it will ultimately be Muslim educators who determine how Islamic schools ought to proceed.

Notes

1. Interestingly, in the Netherlands, the founding of Islamic schools is itself viewed by many as evidence of both integration and emancipation (Spiecker and Steutel, 2001).
2. While the Ministry of Education does impose attainment targets designed to help schools organize their curriculum, and all schools must meet the attainment targets in order to prepare the pupils for the next level of school, schools do have the freedom to decide both the specific content of lessons and the methods of teaching.
3. Partly this may have to do with the larger family size of the average Muslim family relative to comparably situated peers.
4. Indeed, Badawi (2005) found that even poor immigrant parents (in this case, Somali) had comparable levels of education with other, more affluent, parents (Arab or Indo-Pakistani).
5. It is difficult to tell whether, or in what circumstances (witness disturbing developments during the Rushdie affair or the Danish cartoon affair), a devotion to the *ummāh Islamīya* necessarily trumps other forms of citizenship. However, in some schools, there is a much stronger affinity to the culture of origin, thus calling this notion of dual citizenship into question (see Merry and Milligan, in press).
6. It should be noted, however, that in the Netherlands Turkish and Moroccan girls are performing at much higher rates than their male counterparts.

CHAPTER 7

ON THEIR WAY SOMEWHERE:
INTEGRATED BILINGUAL
PALESTINIAN–JEWISH
EDUCATION IN ISRAEL

Zvi Bekerman

Introduction

In 1972, Fr. Bruno Hussar founded the first (and still only) intentionally mixed Palestinian–Jewish village in Israel. The aim was to set an example of coexistence in practice for groups living in what has come to be known as an area of intractable conflict (Bar-Tal, 1999). The village's name is Neveh Shalom/Oasis of Peace (Feuerverger, 2001). In 1984, three years after the first integrated Protestant–Catholic schools opened their doors in Northern Ireland (McGlynn, 2001), an integrated school started functioning in this village in Israel. The school served the local population that, even today, is rather small (totaling sixty families). The school was an exotic educational undertaking serving an unusual mixed community that never developed into an attempt to influence the wider, almost fully segregated educational system in Israel (Nir and Inbar, 2004). Fifteen years later and totally unconnected to these previous developments, two friends—a Palestinian (citizen of Israel) and an American Jew (also citizen of Israel)—started what, at that time, seemed like an impossible (and in the eyes of some consulted experts, undesirable) grassroots movement toward the creation of integrated bilingual schools in Israel. This time the objective was to serve the "regular" population and not those people who already had very clear ideological commitments toward cooperation and coexistence, such as the population in Neveh Shalom. Their determination brought about the creation of the Center for Bilingual Education in Israel, the institutional tool through which the schools under consideration developed, which aims at fostering egalitarian Palestinian–Jewish cooperation in education, mainly through the development of bilingual and multicultural coeducational institutions (Bekerman and Horenczyk, 2004, in press), institutions that at the same time stay committed to the strengthening of each group's identity.

In the interviews I conducted with the codirectors, they reported that they sensed the time had come to correct some of the injustices of the Israeli system. For the Jew, this effort corresponded to his appreciation for the mandates of the Jewish liberal tradition to which he belonged and in which he had been raised, and to his understanding of what a democracy is all about. The Palestinian also spoke about his understanding of democracy and added to this his sense that minorities need to be active in correcting their own circumstances.

For the most part, the sociopolitical context was unsupportive. A long history of conflict and suspicion stood in the way of such an initiative, as did institutional structures that had, for decades, supported segregation. In addition, the lack of any multicultural policy within the Ministry of Education offered little chance for success. Nevertheless, recent historical developments might have created the circumstances necessary to allow for such a "grassroots" initiative to take shape. From their interviews, we learn of the founders' surprise regarding the relative ease with which the initiative evolved. The Misgav Jewish community, situated in the northern part of Israel, was the location of their first attempt back in 1997–1998. During this period of relative calm, following peace expectations upon the signing of the Oslo agreements, perhaps the Jewish community, the minority in this area, sensed the need to strengthen and sustain good relations with the majority Palestinian community. In Misgav, the Jewish population reflected rather liberal political inclinations and a sense that practice needs to support ideology. The Palestinian population reflected no less of an ideological commitment. Additionally, however, they had a deep understanding of the failures apparent in the segregated educational system Israel had created for them and they wished to provide their children with a better alternative (Al-Haj, 1995). These became the forces behind what ultimately allowed for the creation of the first integrated school (kindergarten to K9).

The second school (kindergarten to K10) opened its doors a year later in Jerusalem. During the first year, the school functioned as a section (in first and second grades—ages 6/7) within the Experimental School, in Jerusalem. Within a year, the school moved into a new locality, and in 2007 it inaugurated its own new building recently donated by the Swiss government.

The third school started functioning in 2004 (kindergarten to K5). In a sense, this school is the most revolutionary of the four existing ones, because it is located in a Palestinian Muslim village in the area of Waddi Ara, called Kfar Kara. This is the first time that Jewish parents are sending their children to school in a Palestinian village. The fourth and most recent one opened in the southern city of Beer-Sheba in 2007 (kindergarten—ages 4–5 and 5–6).

The Research

Since 1999, I have been researching the first three of these schools through a long-standing ethnographic effort that included the gathering of information through field observations and videotaping of multiple classroom activities and school events (over 200 hours); interviewing of representatives of all school stakeholders, principals, teachers, parents, and students (over 150 interviews were conducted); and the collection of school documents and curricular materials. These efforts were conducted by my team that at all times included bilingual (Hebrew, Arabic) research assistants from both ethnic groups (Palestinians and Jews). Within the parameters of this chapter, I will only be able to offer some insights into the many educational issues I have identified and analyzed in my work. Those interested in the methodological and descriptive details are encouraged to consult some of my published work (Bekerman, 2003a, 2003b, 2004, 2005; Bekerman and Maoz, 2005; Bekerman and Nir, 2006; Bekerman and Shhadi, 2003). In addition, other scholars who have researched these schools and their work should also be considered (Feueruerger, 1998; Glazier, 2003, 2004).

Studying the schools for the last eight years has brought about many understandings. More importantly, I have come to realize that some of my initial insights, though not necessarily faulted, lacked depth and contextual complexity. One of the most outstanding issues I have had to confront is realizing that the way I have looked at these schools and their functioning in the Israeli society was tinted by macro political formations, such as the nationstate, which fixed my gaze in ways I was not always aware of.

What I have learned always has a Janus character (the double faced mythical figure); it can be understood in more or less positive ways. I do not think that at this point we need to make up our minds. Instead, I think we should stay attentive to this Janus character of the findings and allow it to keep us all—researchers, participants, and practitioners—on our toes. I contend that we must remain alert to the multiple paths new educational initiatives can take, making sure that on the way we sharpen our focus and vision, ensuring that the different potential turns that events can take do not lead us to unwished-for places.

Now, let me say something rather short about the immediate sociopolitical context surrounding the schools, including the main characteristics of their functioning. Like any other state, the State of Israel is a construct developed through a rich imagination (Anderson, 1991) and detailed institutional practices, which include a powerful educational system (Ben-Amos and Bet-El, 1999; Gellner, 1997; Handelman, 1990). Albeit a construct, this does not make the educational system less consequential. The consequences on the Palestinian minority (18 percent of the Israeli population) have been devastating. For the most part, Israel, as an ethnic democracy (Smooha, 1996), has not welcomed the active participation in political, cultural, or social spheres of groups outside of its legitimate invented community of Jews. Palestinians in Israel, though officially offered full rights as citizens, have chronically suffered as a putatively hostile minority with little political representation and a debilitated social, economic, and educational infrastructure (Ghanem, 1998). Indeed, being a Palestinian in Israel is no easy task. The many sociopolitical conflicts that afflict Israel are reflected in its educational system that is divided into separate educational sectors: Nonreligious Jewish, Religious National Jewish, Orthodox Jewish, and Arab. (Palestinians are called Arab in the Israeli hegemonic discourse, thus denying them any national recognition.) All these sectors fall under the umbrella of the Israeli Ministry of Education (Sprinzak et al., 2001). Considering the sociopolitical context just described, the idea of creating Palestinian–Jewish coeducation is, in and of itself, a daring enterprise.

During the academic school year 2007–2008, the four schools comprised a population of about over 900 students with an almost equal representation of Palestinians and Jews in each school. In the future the schools plan to develop into K-1 to K–12 institutions (ages 6 to 17). The schools are recognized as nonreligious schools supported by the Israeli Ministry of Education. Their curriculum is the standard curriculum of the state nonreligious school system, with the difference being that both Hebrew and Arabic are used as languages of instruction. The schools employ what has been characterized as a strong additive bilingual approach, which emphasizes symmetry between both languages in all aspects of instruction (Garcia, 1997).

From their inception, the schools have implemented serious structural changes in the traditional school organization. The initiators' dream was to create an educational context that would practically reflect their vision of equality. Their ideology emphasized the need to sustain symmetry at all organizational and curricular levels. They have been successful in sustaining this goal by securing the services of a well-balanced educational staff, both a Palestinian and a Jewish homeroom teacher in each classroom, Jewish and Palestinian coprincipals in each school, and a well-balanced parents' committee composed of an equal number of Palestinians and Jews. (The homeroom teacher is the central educational figure in the classroom—a leader, organizer, and manager who is the main person to carry on the national educational policy, the educational and teaching plan for the school, the link of the home to the school, and promote student development.) Just watching this heavily populated and well-balanced corps of practitioners, administrators, and parents function in a rather harmonious way is already a feast when considering the very basic structural inequalities that characterize the Israeli society described above. A setting that, in its very structure, assumes the need for equality must be doing something right.

Nevertheless, we should remember that the schools are supported by the Ministry of Education as regular schools, and regular schools cannot afford two homeroom teachers and coprincipals. Thus, the very basic structural changes implemented are in need of extra budgets for their existence. These extra budgets are raised partially by the NGO (nongovernmental organization) behind the initiative and partially by parents paying a rather high tuition (over $US1000 per year) when compared to the traditional cost of public education in Israel (which, depending on the area in which a school is situated, ranges from $US100 to $US400). Given the tuition costs, it is not surprising that, until now, the schools have attracted a rather well-to-do population of academics, professionals, and independent merchants able to afford these costs. These budgetary constraints limit the growth of the educational initiative. There is little indication that the Israeli Ministry of Education will change its policy and decide to support these schools according to their expressed ideological need, and thus the initiative is at risk of becoming one that will serve only certain elites in Israeli society.

Regarding their educational practices, the schools seem to be faring well when their educational achievements are compared to other, segregated educational institutions. While not all assessments required by the state have been carried out, it seems that in some cases, unsurprisingly, Palestinian pupils in the bilingual schools are doing better than those in segregated Palestinian schools. This relative success is of utmost importance to the future of the initiative, since, as we mentioned earlier, in spite of the political ideological inclinations of the populations participating in the schools, as members of an upper-middle socioeconomic sector of society, education is mostly seen as a means of mobility in a world going global.

Like many bilingual programs, the bilingual schools studied suffer from somewhat contradictory practices, perspectives, and expectations in relation to their goals. During the first years of the schools' functioning, teachers experimented with ways to prevent any of the languages used at school from becoming segregated or compartmentalized into any specific discipline or time slot, while continuously trying to assure equal use of Arabic and Hebrew in class practice. This was a laudable and demanding endeavor considering that, to a large extent, such a bilingual strategy had to be created from scratch. Moreover, in Israeli society, Palestinians are, or easily become, bilingual, thus allowing the school to hire Palestinian teachers who dominate both languages. On the other hand, Jews have been raised in a monolingual (Hebrew), monocultural (Jewish) context (Spolsky, 1997). Arabic (mostly considered the language of the enemy) is rarely one of the choices as a second language, making it almost impossible to hire Jewish bilingual teachers. The serious efforts invested by the entire staff to sustain full symmetry through the implementation of bilingual educational practices seemed to fail already during the first years of the schools' activities. Even when the language policy shifted toward an even stronger support of Arabic, the introduction of English (starting in first grade) and Israel's homogenizing policies and context rendered in all four schools the bilingual efforts mostly ineffective with regard to the Jewish population at school (Bekerman, 2005). In spite of multiple new curricular changes, which have been recently implemented, it is difficult to believe much progress can be made (Amara, 2005).

While teachers and the NGO see this as a serious obstacle to achieving their declared goals, both Jewish and Palestinian parents seem less worried. Teachers were conscious that much of what went on was still biased toward the Hebrew language. All teachers' interactions we observed and recorded were conducted in Hebrew, and Hebrew was also the prevailing language in all staff meetings and training sessions, as well as in the meetings of parents and of the steering committee. In their interviews, teachers and students who related to the issue of language use indicated that interactions in class and recess between children of different national groups were conducted in Hebrew. This pattern also emerged clearly from all our videotaped and written observations. As for parents, Jewish parents, for example, support bilingualism as long as it does not harm educational excellence. They seemed satisfied with an educational initiative that allows them to substantiate their liberal positions and to offer their children cultural understanding and sensitivity toward the "other." Palestinian parents seemed to be after the best education available given the present Israeli sociopolitical context. As apparent from the interviews we conducted, Israel's present sociopolitical conditions make it almost impossible for Palestinian parents to dream about a soon-to-arrive top-down multicultural multilingual policy and, given their mobility aspirations for their children, they prefer an English lingua franca and high Hebrew literacy while secure that acquisition of Arabic can be attained within the family and community context.

English started being taught the second year of our study. English successfully captured the goodwill of both Palestinian and Jewish children and their parents. With the appearance of English in school, we have come to hear more Jewish parents' and children's voices questioning, if not the absolute need to study Arabic, the amount of time invested in it. In the larger context, where Hebrew is the dominant language and English potentially offers a free pass into a global reality, Arabic risks being undermined completely. Context might be too powerful to be overcome even by the most well-intentioned bilingual educators. Still, it also becomes apparent from our interviews with Palestinian parents that the present policy of affirmative action favoring Arabic, together with the schools' legitimization of Palestinian identity and symbols, allows them to be confident in their decision to send their children to an educational setting that otherwise would be ideologically difficult to defend from their perspective as a minority. During the interviews, more than three-quarters of the Palestinian parents mentioned that they were able to respond to criticisms voiced by relatives and neighbors by pointing to these language policies and their implementation.

Paradoxically, when considering the above, strengthening the Arabic language in the school curriculum could mean working against the Palestinian minority needs and expectations for successful education. Parents and students were clear that strong Hebrew literacy and a good domain of the English language would secure the success of their children in higher education, which in Israel is solely conducted in Hebrew (with relatively long English bibliographies). Parents were unambiguous about their children reaching a high level of Hebrew and English literacy so as to ensure that their children would not suffer as they did when entering the Israeli universities. In 2002, the NGO changed its official name from the Center for Bilingual Education in Israel to the Center for Arab Jewish Education in Israel. Though never officially acknowledging it, this change might reflect the institution's understanding that bilingualism is not the main choice or expectation of their population and that in the present Israeli sociopolitical context, where Arabic carries little cultural capital in the market place, it might not be achievable. Indeed, language communicates, but in our contexts it mostly does identity politics.

Multicultural curricular efforts seem to be fairing much better than the bilingual ones (Bekerman, 2002, 2004). In our interviews, parents emphasized that culture and religion are the areas where mutual understanding can help bridge the gaps that separate both populations in Israel and can contribute to peaceful coexistence. Parents stressed the importance they see in getting their children to know and understand the others' culture better. They believed that this is well on its way to being achieved successfully in the schools. Although throughout the year cultural and religious issues are raised and discussed, for the most part they become salient during special events such as school trips (for example, visits to a nearby synagogue, mosque, and church), or festive events at school such as the celebration of Hanukkah, Id el Fitter, and Christmas, which allow for broad expressions of solidarity and mutual understanding.

What becomes apparent during these celebrations is that they carry a strong religious emphasis. Paradoxically, it could be said that religious aspects are disproportionately emphasized in these schools, when considering that for the most part the parents belong to the secular segments of Israeli society. In the schools, all of the Jewish parents are secular and the Muslim population, though more traditionalist, is mostly nonreligious. It is important to note that in all three schools special time has been allotted to religious studies, which are conducted, for the most part, separately for the Jewish, Muslim, and Christian populations. In some interviews, Jewish parents expressed some concerns and ambivalence about this religious emphasis, but, at the same time, they seemed to find solace in the religious underpinnings given their mostly unarticulated fear of the erosion of their children's Jewish identity as a result of their participation in an integrated bilingual program. This attitude seemed to be shared among Jewish teachers, who expressed their need to emphasize knowledge of Jewish traditions as an antidote to the perceived superficiality of secular Jewish identity and its possible weakening within an integrated environment. Within this chosen educational sphere, and the threats it offers, teachers and parents seem to have no choice but to support the use of strategies that will sustain the particular cultural traditions of the groups involved. Trying to make sure their children do not lose their sense of group belonging is of utmost importance for both Jewish and Palestinian parents (Bekerman and Maoz, 2005).

Ironically, an educational sphere meant to soothe national conflicts finds itself emphasizing cultural differences. Even if, for a moment, we would find this chosen direction desirable, as we could indeed, we should ask if that which is offered in terms of religious or cultural artifacts does indeed do justice to cultural traditions that are over one thousand years old. I doubt the answer could be positive. I doubt if rather shallow representations of Christmas, Hanukkah or Id el Fitter, or Rosh Hashanah and Ramadan, or hummus, pita, and latkes can do justice to ancient and revered traditions. These traditions have been responsible for the development of worldviews that, throughout the ages, have produced profound literary and scientific products that until today feed the imagination of a thinking humanity. The building bricks of these civilizations cannot be found in the school curriculum. Rather, shadows of them in the shape of truncated holy texts or cuisine recipes make their appearance in the school scene. Unfortunately, for the most part the multicultural strategies adopted can be easily criticized for their essentialization of culture (the marking of multiple educational activities and events as related to ethnic categories) and its reduction to some reified folkloristic perspective (Bekerman and Maoz, 2005).

The ethnographic data suggest that coping with issues of national identity has become the ultimate educational challenge for parents and educational staff alike in the bilingual schools. National issues are easily compartmentalized into a rather short and discrete period in the school year: those dates corresponding in the Jewish Israeli calendar to the events of Memorial (in commemoration of the Israeli soldiers who lost their lives in all Israeli wars) and Independence Day and in the Palestinian calendar to the Day of the Naqbe (the Catastrophe—the tragedy that overcame the Palestinian population during the 1948 war, which brought about the creation of the State of Israel). The schools hold separate short ceremonies for each national group on these national remembrance dates.

The progression of these events is immensely influenced by political developments in the area. During the first year (1998–1999) of our study, when Israel was still under the optimism following the Oslo agreements, these events progressed rather smoothly. During the second year, they were strongly influenced by the tense and violent political climate in the outside world, and their preparation was carried out under a growing sense of suspicion on the part of both the national groups represented at the three schools. Yom Ha'Adama (Day of the Land, commemorating Jewish confiscation of Palestinian land) also turned out to be a difficult event, with fifteen Jewish parents contacting the principal in the Jerusalem school questioning the unfavorable ways in which Jews had been represented at a school exhibit and the possible effect of the texts invoked in the ceremony on their children's perceptions of themselves as Jews.

The Jewish group appears to feel highly threatened during these commemorative events. This may be due to its status as a majority, which, for the most part, enables its members to take their identity for granted, given Jewish hegemonic power in Israel. The Jews at the schools clearly represent the politically liberal center–left segments of the Israeli society. In all the parents' meetings dealing with cultural and national issues, as well as in most of the interviews, they expressed openness to the needs of the Palestinian population at school and a willingness to grant expression to their identity. However, the forms of Palestinian expression do not always seem to fall within the limits of legitimate Palestinian expression as delineated by the liberal Jews. For many of them, Israeli Palestinian cultural and religious expression in school is legitimate, but national identification with the Palestinian Authority is not always welcomed (Bekerman and Horenczyk, 2004).

Palestinians, who as a minority have become used to subduing the expression of their national consciousness within the State since its creation, approach the school with restrained expectations regarding their national needs. They clearly expect not to have to hide their national cultural identities but seem to be ready to apply the many mechanisms they have developed over long years of repression, such as not attending school on national Jewish days or just quietly reinterpreting ritual events when compelled to participate in them. In all the interviews, Palestinian parents and the Palestinian educational staff express appreciation for the more liberal approach offered by the school initiative and for what they perceive to be the honest openness of the Jews toward their needs.

Data gathered primarily during the parents' meetings and during the steering committees clearly show growing expectations among the Palestinian parents to give what they see as proper expression to their national identity. Most of the Jewish parents react with surprise, and at times with a sense of fear. At first they were happy to consider themselves true liberals, willing to open up their doors to those oppressed; later they had to face their own misconceptions and fears (as with the Land Day commemorations mentioned earlier). Palestinian parents, for the most part, refrained from openly criticizing these views. When considering the very few options available to their children in the present Israeli school system, they prefer to try to work things out quietly, usually in ways that allow them to find adequate solutions to these problems without betraying their own feelings and symbolic needs. Considering current and historical circumstances, any step further can truly be conceived of as a triumph. And there were such triumphs. The format agreed upon in the bilingual schools, which includes separate Naqbe and Soldier Remembrance Day ceremonies as well as Land Day commemorations, are outstanding examples of progress in Israel, so much so that mainstream schools are trying to learn from the bilingual experience.

Lastly, we will mention some of our findings regarding children's interactions. At the beginning of our research during the first years of the school initiative, nobody seemed to be really happy when considering social interactions among the students of the two main ethnic-national groups represented in the school. Not being really happy does not mean that stakeholders were upset. It just means that they expected more. In general, we found that intergroup interaction in classrooms was greater than during recess and in other unstructured periods. Recurring patterns observed during the recess periods showed Jewish and Palestinian children playing more in separate groups than in mixed ones. Most of the children were aware of this pattern, but they seemed not to regard it as a problem, or they explained it in terms of personal preferences (for example, such and such children don't like to play soccer). In class, children worked together in cross-national teams and assisted each other with different assignments. When at home, however, intergroup visits are rare. Exceptions do exist, but in general the educational staff and parents kept questioning why more gains had not been made in this area. Recent years present a much more integrated picture, partially explained by the fact that children have been together for a much longer period of time and naturally have become friendlier, but still some of the patterns uncovered in the past can still be seen today (ethnic grouping and play during breaks). Attempts to explain the situation are varied. Some point to the distance between the settlements in which the children live, or the segregated areas inhabited by the groups; others point to cultural differences regarding the protocol of social interaction; still others mention that language gaps are a barrier to social communication outside of class.

Over time, it has become more and more apparent to us that the younger the children are the less attentive they are to ethnic-national differences. Our observations in kindergarten classes made it very clear that children, regardless of ethnic affiliation, felt the same affinity to both the Palestinian and the Jewish teachers. Moreover, throughout our observations we never recorded any incidents that could suggest ethnic/religious tensions among the children, not even when small disagreements occurred, as in the case of a discussion about who had picked up a game first. Even when theological issues were discussed by young children, the alignment around an argument could not be predicted on ethnic/religious lines. The fact that ethnic/religious differences seem to play no part in the children's daily interactions should not be understood as meaning they have no consciousness of their ethnic/religious affiliation. They do have such awareness and they even play with their identities (children were observed adopting a different ethnic identity during playtime). Still, this clear knowledge of ethnic/religious affiliation never seemed to become an obstacle or to be much attended to in the daily activities.

Our records gave us the general impression that the kindergarten children are a bit (just a bit) more inclined to affiliate (talk and play) according to ethnic lines. Additionally, this tendency seemed to grow from first to third grade, creating the impression that ethnic lines become more salient as the children progress in age within the school context. We should be very careful when making this statement since any attempt to measure these inclinations has become a methodological nightmare and a task we believe cannot be fully achieved, particularly if we are serious about taking into consideration contextual circumstances. Still, we can say, with a fair amount of conviction, that there is always a numerical asymmetry in the groups when playing or chatting among themselves in informal settings (activities not directed by teachers). We emphasize the informal activities because, as we previously mentioned, during formal activities the teachers make sure that all groups are regularly mixed. For example, in the classroom the children sit in a way that ensures that each hexagonal table is populated by mixed groups. For the most part, these groups are uni-national ones with children from the other group joining in smaller numbers. A second asymmetry regarding these mixed groups is that without exception they talk in Hebrew, something deemed to be necessary by all involved since they are well aware that the Jewish children lack Arabic proficiency. Still, this asymmetry was never mentioned as detrimental to the social relations among the children, who seem to accept it as a matter of fact.

Our records from regular classroom activity point in the direction of a strong integration among the children, which seems not to be organized along ethnic lines. If at all organized, this organization seems to work along gender lines. We have recorded many episodes in which a type of transgression of accepted behavior is enacted by a child belonging to an ethnic group toward one in the other group; what is important to mention is that never in such events did we register any recriminations that, in any way, included noticing the ethnic background of the transgressor as if this would be totally irrelevant.

Paradoxically, observing the activities of the teachers we find a rather dissimilar picture. Over time, teachers have developed strong personal ties with each other regardless of ethnic background. Nevertheless, they are adults raised in segregated schools and asymmetrical relations of power (which always benefited Jews). In addition, they sense by themselves and through outside pressures the tensions of, on one hand, being willing to support tolerance and peace efforts and, on the other hand, the great success of hegemonic powers in the shape of parents, the ministry authorities (supervisors, official curricula), and the NGO behind the educational initiative, which implies that such initiative is possible only when the group identity of the participants is saved and salient. It is mostly the teachers who remain, at times, in the crossfire between contradicting goals. Negotiating a curriculum that works toward inclusion and differentiation, though possible, is clearly not easy. The encouragement of strong social relations are supposed to be high on the agenda, but their possible outcome in the future—interethnic marriage—is not a preferred option. The development of a curriculum that fully and openly acknowledges the dichotomous historical perspectives of the groups involved is considered a necessary achievement, but throughout it is only the teachers (also as regular citizens) who are the ones who need to confront in the class frontlines the differing perspectives on what this exactly means in the eyes of the multiple stakeholders who hold the key to the potential future development of such an initiative—the Ministry of Education authorities, parents, and the children themselves. Inclusion, acknowledgment, and tolerance are also high on the agenda, and a multicultural curriculum is expected to serve these central aims. However, not only does such a curriculum not exist but, if it were to be found, its implementation would need to restrict the crossing of ethnic-national cultural boundaries by children even if, by chance, they found the contents challenging or enchanting. All in all, teachers are the ones who take in all the pressures that follow from a grassroots initiative that is far from having been able to fully and rationally articulate its goals. The schools are, indeed, an alternative for something. What has not yet been well articulated is what that alternative is all about.

Theoretical Issues

We mentioned in our opening remarks that throughout the research process we learned significantly about our own limitations as researchers and that much work has been invested in uncovering the contextual issues that limit our gaze when trying to make sense of this outstanding educational initiative. I want to close this chapter by addressing some of the theoretical issues that both need to be further researched and critically approached to make our analysis toward understanding better fitted to confront the many questions such an initiative raises regarding issues of educational reform.

Given Israel's present realities, at this point the integrated schools present a partially egalitarian option unheard off in the surrounding context and as such serve as an example of how "things" could look even in deeply conflicted societies. In general, we might not want to support a large bottom-up reform project that serves only to further the mobility of those already mobile, even if these are mixed Palestinian/Jewish groups. However, as a small project it might contribute to a change of rhetoric, which might subsequently inspire others to follow. The schools also raise multiple questions regarding the theoretical and practical potentials of multicultural and bilingual educational efforts as a conduit toward the implementation of a just society. Though not conclusive, our research shows some of the limitations of multicultural and bilingual efforts. If uncritical, multiculturalism seems to fall prey to essentialist and reified perspectives that position educational practices at the service of political agendas within the realm of the nation state and in a world where national boundaries no longer represent clear-cut national identities. In doing so, multiculturalism supports hegemonic powers, and though it allows for certain sectors of society to benefit it does little in terms of helping change the rules of the power game. Similarly, bilingualism seems to suffer from contradictory strategies in its implementation. At present, the bilingual project seems also primarily to serve political agendas benefiting mostly the needs of the nation-state. At times, these agendas strive sincerely to promote the interests of minority groups and confront mainstream hegemonies. At other times, such agendas are just paying lip service to political correctness. For whatever reasons, these initiatives seem, in the best of cases, not to be attaining their goals and, in the worst of cases, oblivious to the reasons for their failure.

As for the benefits of integration, research on intergroup contact shows that such contact does generally promote intergroup acceptance, especially when appropriate conditions for the contact are being met (Thomas F. Pettigrew, 1998; T. F. Pettigrew and Tropp, 2000). Much of this research stood as the basis for the Brown *v.* Board of Education of Topeka, Kansas (Brown v. Board of Education, 1954), the Supreme Court decision that ordered in 1954 the desegregation of public schools throughout the United States—a decision considered by some to be the defining U.S. legal decision of the twentieth century.

Still, views are divergent as to whether the Brown decision has fulfilled its promise. Those discouraged point out that even today, though not state-imposed, many school districts remain segregated by race (Guinier and Torres, 2002). Even in integrated settings, students of color find themselves re-segregated through academic tracking or social interactional boundaries (Tatum, 1997). They find themselves disproportionately represented in poorly funded schools (Glickstein, 1996), and for those students who do make it to college, their drop out rate is much higher than the rate for white populations (Bowen and Bok, 1998; Steele, 1997). The optimists point to the tremendous growth of students of color in the percentage of college graduates (Bowen and Bok, 1998) and to the fact that many educational institutions are involved in developing strategies that will further support the development of students of color through a variety of tools such as affirmative action and a multicultural curriculum. These steps, as well as others, all try to readdress racial disparities in order to attain the primary goal of Brown's decision to offer equal access to educational opportunities (Wells, 1996; Wells and Jones, 1998).

While the results of Brown *v.* Board of Education may not be conclusive, there is a growing impression that scholars are becoming more and more critical, if not about the essentials of the Brown decision, then about its implementation. This criticism is so strong that some have recently argued that racial disparities would have been better served if Brown had upheld the Plessy v. Ferguson decision of 1896, which provided for "separate but equal" settings. As evident in today's reality, educational settings are still separate but unequal (Bell, 2001).

When considering the above, it becomes apparent that desegregation directed toward integration is not necessarily the correct answer to the problems of a diverse and conflicted society. The United States' example shows that even when initiated by the state, integration is hard to achieve, difficult to sustain, and even harder to maintain (Gallagher, 2007). Though there seems to be value in the use of legislation to overcome legalized segregation, it is doubtful whether legal steps can overcome custom and practice. Localized behavioral changes imposed by law might not be able to influence the wider context. The powerful always seem to find ways to evade legally imposed integration.

A possible solution to this conundrum is for integrated schools to be seen not just as meeting places for students from different backgrounds but as spheres in which the curriculum and the school ethos deal directly with areas of diversity while being committed to addressing cognitive and affective issues over behavioral ones.

But such an approach is yet to show if it can be successful in the integrated schools such as the ones initiated in Northern Ireland and Israel. We could easily find fault with the teachers and parents involved in our program. We could blame them for consciously or unconsciously conveying negative messages about the groups involved in the conflict in spite of their overt efforts to create a school environment and a curriculum that represents a balanced multicultural effort. But this would be clutching at straws. If placed anywhere, the "blame" should more accurately be placed on an adaptive, wider, sociopolitical system in which minorities carry little symbolic power. There are no universal multicultural approaches that can be offered independent of sociopolitical contexts. Every multicultural endeavor involves new imaginings and difficult, hazardous work.

When we consider the proliferation of conflict worldwide, the importance of supporting those undertaking these worthwhile educational experiments in such challenging contexts is self-evident. The benefits of developing and sharing practices about learning with the other are also apparent. This chapter suggests that there is a need for those involved in integrated educational efforts to look at a combination of theoretical perspectives for guidance on their policy needs and for academics to support them on their journey. The integrated schools challenge our theoretical imagination for they compel us to consider specific individual experiences as well as individuals and their group affiliation. Cross-fertilization between present paradigmatic perspectives, considering how contact and acculturation theoretical perspectives can inform each other, and their better adaptation to the need to account for individual behaviors in all their complex details and potential outcomes might be the direction to take if we want to assist integrated bilingual education. However, we would do well to remember that, even as education can reflect and support change, education by itself cannot bring about what is truly needed, the structural changes that underpin equity.

Notes

The research project was funded by grants from the Ford, Spencer, and the Bernard Van Leer Foundations. I'm also indebted to Julia Schlam for editing the manuscript and offering throughout insightful, critical comments and suggestions.

Further Sources of Information

Bekerman, Z. (2003a) "Never Free of Suspicion." *Cultural Studies Critical Methodologies* 3/2: 136–147.
Bekerman, Z. (2003b) "Reshaping Conflict through School Ceremonial Events in Israeli Palestinian–Jewish Co-Education." *Anthropology and Education Quarterly* 34/2: 205–224.
Bekerman, Z. (2004) "Multicultural Approaches and Options in Conflict-Ridden Areas: Bilingual Palestinian–Jewish Education in Israel." *Teachers College Record* 106/3: 574–610.
Bekerman, Z. (2005) "Complex Contexts and Ideologies: Bilingual Education in Conflict-Ridden areas." *Journal of Language Identity and Education* 4/1: 1–20.
Bekerman, Z. and Maoz, I. (2005) "Troubles with Identity: Obstacles to Coexistence Education in Conflict-Ridden Societies." *Identity* 5/4: 341–358.
Bekerman, Z. and Nir, A. (2006) "Opportunities and Challenges of Integrated Education in Conflict-Ridden Societies: The Case of Palestinian–Jewish Schools in Israel." *Childhood Education* 82/6: 324–333.
Bekerman, Z. and Shhadi, N. (2003) "Palestinian–Jewish Bilingual Education in Israel: Its Influence on School Students." *Journal of Multilingual and Multicultural Development* 24/6: 473–484.
Feueruerger, G. (1998) "Neve Shalom/Wahat Al-Salam: A Jewish-Arab school for peace." *Teachers College Record* 99: 692–730.
Glazier, J. A. (2003) "Developing Cultural Fluency: Arab and Jewish Students Engaging in One Another's Company." *Harvard Educational Review* 73/2: 141–163.
Glazier, J. A. (2004) "Collaborating With the "Other": Arab and Jewish Teachers Teaching in Each Other's Company." *Teachers College Record*. 106/3: 611–633.

Chapter 8
"Alternative" Māori Education? Talking Back/Talking Through Hegemonic Sites of Power

Hine Waitere and Marian Court

To talk about indigenous Māori educational initiatives
is to talk about Māori struggles to endure as
fourth world peoples within first world contexts, to endure
as Māori.

To talk about alternative Māori educational initiatives
in English,[1] is to risk
talking about difference in ways that
inadvertently re-inscribe the pathologies of being different,
of being "other."

Risk—yet opportunity
to offer some images, some stories—
to conjure some reflections and questions...
Hei aha tēnei? What is this—Māori education as alternative?

To begin—some images...

Ngā Manu Kōrero[2]
(Māori Speech Competitions)

In hallowed school halls, rows of
  tired
  sometimes splintered
  always uncomfortable
  and unforgiving
  forms
  coldly delineate formal protocols,
  limiting boundaries of possibility....
  deflecting challenge,
  resisting change...

*But here, now
assaulted doors succumb to
    low riders in Grey Shorts
      kilted skirts,
       l o n g trousers,
            blazers,
         monogrammed school jerseys,
              branded sweat shirts,
      pleated pinafores,
A-lined skirts,
      number ones and travelling trackies spilling across the
breached divide—
    down the aisle students and supporters surge
peeling left, tracking right successive waves of bodies spilling into
contested seats.*

*A marauding mass, tightly packed, moving as one. Old bodies,
              young vessels,
          gumboots
                  snotty noses
           designer coats
   competitors,
       supporters and
          spectators
                  all jockey for space—
              they sit on seats, stand in aisles
         and lean against walls
    in the co-opted school hall.*

*Scraping metal chafes wooden floors,
      creaking seats groan submission to shifting bodies.
    There's a shuffle of regulation shoes and
    the drone of expectant whispers oozing
      anxiety,
          quiet confidence,
              self-doubt and nervousness…
adrenalin rushing through sweaty palms—
        as the latest group of manu kōrero
        anchor to their entourages.*

*They are different. Different schools. Different programmes. Different levels of linguistic competency. Different tribal affiliations.
Different.
Different.
Different.
Yet they are all the same.
Each competitor must claim Māori whakapapa[3] or competence in te reo Māori[4], to be eligible to compete at Ngā Manu Kōrero.*

*A rising wave of silence subdues the last resistant whisperer, quietens shuffling
shoes and stills the heaving sea of bodies
into a montage of green and
         tints of red,
       shades of blue,
       hues of yellow,
         pockets of brown,
       tinges of beige—
colours demarcating boundaries that separate yet unify Māori and
students of Māori encased within.
Tihei Mauri Ora!*[6]

Some Questions and Reflections

Ngā Manu Kōrero, these Māori speech competitions and speech-making might seem strange, yet familiar, to these students' tūpuna,[6] the kuia,[7] and koroua[8] who have preceded them. Those ancestors, many of them well-versed in sophisticated, solemn, mischievous, funny, wise, and political oratory, might justifiably ask, *he aha tēnei?* What is this? They might query, what makes these competitions alternative, or conversely, how do Ngā Manu Kōrero relate to tradition and/or customary practice? What connection do these speech competitions have to whaikōrero,[9] with whakataukī,[10] karangā,[11] oriori,[12] pātere,[13] waiata,[14] indeed with Mātauranga[15] Māori and customary educative processes and practices?

You (our readers) might also be asking, he aha tēnei? What is this? What does Ngā Manu Kōrero (a state-initiated secondary school speech competition designed to promote a greater command and fluency in the use of spoken English) have to do with alternative education or with the revitalization of an indigenous language? Others may justifiably ask, what does it mean when one evokes the notion Māori education?

The last question is one we (one Māori and one Pākeha[16] academic) continue to ask ourselves and our students in the university where we work. In a number of our courses focused on Māori Education, preservice and postgraduate teacher education students are asked, what do we mean when we talk about Māori education? In an undergraduate class,[17] team-developed, taught, and modified by a group of Māori educators, students initially bewildered by the fundamental question would invariably raise subsidiary questions. Probingly, some ask whether Māori education is about education that is focused on Māori content, or is it educational provision for Māori children? The quick response is that politically, at least for Māori, the push is for both of these things. Students are further asked to consider though, what Māori education might look like on the occasions when we are talking about content and we overlay its delivery with the identity

	Māori teachers	Pākeha Teachers	Immigrant Teachers
Māori Pupils			
Pākeha Pupils			
Immigrant Pupils			

Figure 8.1 He aha tēnei: What is This Thing Called Māori Education?

of the teacher, and/or on the diverse range of students who increasingly inhabit our classrooms. Throughout their study in these courses, we ask the students to work individually and together to think about what Māori education might look like in the intersections constituted in the grid represented in figure 8.1.

For example, in the first box the students consider whether Māori education is about Māori teachers teaching Māori children about things Māori. If so, what does this mean for them, as student teachers who are mostly Pākeha, female, and middle class? Or (to move down the grid), is Māori education about Māori teachers teaching Pākeha, or perhaps migrant pupils, about things Māori? Or placing themselves in the centre, is it about Pākeha teachers teaching Māori pupils about things Māori? Up to this point, discussions usually focus on bicultural relations (as between Māori as tangata whenua[18] and Pākeha as Treaty partners). But when we further complicate the question he aha tēnei o Mātauranga Māori in contemporary multicultural contexts, the questions must also include, what does Māori education look like or feel like for immigrant teachers from diverse linguistic, cultural, ethnic, and religious contexts where the political environment is structured around biculturalism as being the primary relationship that precedes multiculturalism? Whatever our responses, what are the implications for the programs and practices we create and engage in as teachers?

Thus we further explore with the students whether these questions have any bearing on what is taught (curricula issues), how it is taught (pedagogical issues), or how we know students have learnt what we intended them to learn (assessment issues). Invariably, when we talk about Māori education we are evoking all of these things. We are talking about the politics of knowledge (what is worth teaching and who gets to decide); we are talking about the politics of place (the geopolitical context in which education is enacted); and we are talking about the interrelational nature of both these things. At these intersections the fabric of colonial narratives begins to unravel for a number of students.

As they continue in their study of, and work in, teaching and learning, the students are encouraged also to keep thinking, as we do, about the impacts on Māori peoples as the tangata whenua in Aotearoa, when their tribal homelands, culture, language became encased in a colonizing nation state. These are some of the geopolitical complexities that constitute indigenous education in a Fourth-World[19] context. For the teacher educators teaching these courses (whether it be teaching students who are studying in general-stream, bilingual, or immersion Māori programmes), the challenge is to create a counter-hegemonic space. That is, a place where they learn that self-determination for Māori is not a gesture of charity, the benevolent provision of an alternative option: it is an act of justice, the realization of a democratic ideal enshrined in the Treaty of Waitangi that centrally locates Māori within the mainstream. The Treaty of Waitangi as the founding document for New Zealand's bicultural heritage recognized the preexisting sovereignty of Māori.[20] Consequently, the signing of the Treaty in 1840 did not confer but affirmed the right of Māori people to retain their way of life and language, while also according them equal status with their treaty partner with full rights of British citizens. In subsequent decades, while the rights afforded the British Crown in the Treaty were unscrupulously upheld, the Crown's responsibilities to Māori were, more often than not, overlooked.[21] Worse, assimilatory state policies and large-scale land purchases and confiscations were enacted along with discriminatory educational policies and practices. Consequently, Māori have struggled against political marginalization and economic disadvantages to ensure the survival of their language and culture and to achieve "official" status as the tangata whenua.

Given all these complexities, we wondered how we, within one short chapter, could even begin to capture some of the diversity of Māori peoples' experiences in, struggles over, and creative recreations of Māori education within the context of colonization. Further, how could we adequately contribute to this book's aims to build mutual sharing, exchange of ideas, and learning between different groups, given the nature of challenges that rest not only on what is being talked about, but on who is doing the talking? Who is centered in the posing of questions about alternatives? Who is "being represented," by whom, for whom, under what conditions and with what effect? Should we refuse the invitation, maintain silence?

A Story

As we pondered these questions, Hine said, "Last week I judged at the Manu Kōrero regional competitions. I sat at the front of the school hall, watching students literally spill into and fill the space. As they did it occurred to me that I was witnessing a transformation of space—a place often described as toxic. As I watched groups arrive from different places, representing different programs—Kura Kaupapa Māori,[22] rumaki,[23] bilingual and general-stream programs—all gathered together, waiting for the pōwhiri[24] to begin—it struck me that, more than any other occasion (whether it be cultural or sporting), standing before me were representatives from the full range of educational options that constitute Māori education. All students, Māori and Pākeha alike, were there for the same purpose: to support, listen, or to speak—about matters Māori, in Māori, and in English, for, to, through, and with Māori. Competitors, all supported by peers and whānau,[25] crowded into a general-stream school hall, one that was being transformed, governed by Māori protocols and practices." Hine paused and Marian responded, "Last night I was at my granddaughter's primary school speech competition. The 'hall' was small, more like a big classroom, with the kids who were to speak sitting nervously on one side and the mums and some dads and a few granddads and grandmas lined up as audience. It was great! Full of anticipation for all of us—and quiet excitement. There were two young Māori students, a boy and a girl, among the twelve or so Pākeha contestants, all winners of their own class competitions. The speeches were all in English though and none of the speakers focused on things Māori—but they gave us fresh insights into issues of the environment, women's suffrage, favorite things, a magic moment..."

Across our thoughts about these similar yet different competitions and celebrations of young students' oratory skill, the word "alternative" skipped. In these contexts, what does "alternative" Māori education mean, we asked ourselves?

Māori Education as Alternative/Alter-Native/(Alter)Native

A surface reading of these speech competition vignettes might be merely that Māori education is *an option*, providing an *alternative* or another choice within mainstream. And given that Māori education has come to mean a myriad of things (as seen in the discussion of figure 8.1), certainly this is one interpretation. Reading within frames of sociocultural power relations, however, poses more significant considerations that ask what it means for Māori people to live and work within institutions and rules not of their own making. Here the notion of Māori education as an alternative refracts in curious ways, appearing as *abnormal*, juxtaposed against invisible and unspoken norms, appearing as peripheral within the centered hegemonic mainstream education. Within this frame, the thrust for Māori education can quickly become *alter–native (alter-the-native)* education for Māori.

Successive waves of educational assimilative policies and practices attest to the number of attempts to establish that version of educational alternative for Māori. In fact, Ngā Manu Kōrero emerged from this ideological space, as we will show. Should the question be reversed, then, as in, what is the education that is *(alter)native—in opposition to alter-native* assimilatory education? If this sense is taken, should we be talking about Native schools (established in Aotearoa New Zealand in 1858)? Or the ongoing struggles for control that led to the development of contemporary Kura Kaupapa Māori initiatives? Should we in this chapter be reexamining how the former sought to "save" a "noble, but savage race," or unpacking how the latter are embedded now in struggles over standards and accountability regulations imposed on kura as a trade-off for winning state funding under the provisions for "Special Character Schools" (Education Act 1989)? Here we would be exploring what has emerged as Māori have moved out to create their own parallel systems,[26] challenging readings of *alternative* that simplistically equate *other* or *different* in ways that inadvertently reinscribe the very pathologies of difference that they attempt to negate. However, perhaps the most radical and thus fraught alternative for Māori education, is to not only reject readings of it as *alternative* (an option) or *alter-native* (assimilated), but to place it as *(alter)native* mainstream. This placing rejects ongoing positionings of Māori people as physiologically, ideologically, or philosophically reluctant "tourists" in an educational system not of their own making: it accepts and promotes the legitimacy and centrality of te ao Māori as mainstream education in Aotearoa New Zealand.

In the rest of this chapter, we look at Māori education through each of these sociocultural lenses of *alternative, alter-native, (alter)native* to/within a mainstream education system/philosophy/approach. In doing so, we are hoping to signpost for ourselves, as much as for anyone else, some of the potholes and pitfalls of locating ourselves outside the manifold iterations that historically and currently constitute Māori education in the compulsory sector in Aotearoa New Zealand. As illustrated in our discussion of figure 8.1, Māori education cannot be seen as one definitive program or practice aimed at a "brown" target group located simply on the periphery of a "white" centered state education system. What has been seen as a brown/white divide of Māori/Pākeha relations is not a neat dichotomy: since first contact, brown and white threads have chafed, entwined, and disentangled. Thus, this is a complicated terrain of uneven tensions and contests, such as suggested in Bhabha's (1994) notion of the "third space"— relational spaces within dynamic cultures that are in constant interplay. Furthermore, thinking about Māori education within a centre/periphery oppositional framework (that is, as an alternative *to* the mainstream) risks merely recapitulating and re-embedding the white Anglo-European norm that is valued in that dichotomy. Yet Māori struggles, resistance to and co-option of normalizing tactics of the state as these have been played out in the field of education, must not be ignored or downplayed. There have been multiple "fronts" in Māori peoples' struggles in their attempt to be mainstream. The following discussions illustrate one "front," the work of problematising the Western pursuit and "taking" of indigenous intellectual and cultural knowledges in order to further the imperial colonizing interests. This work has included tracing and recording Māori resistance to and deconstruction of those practices (Smith, L., 1999, 2–3). We interweave into our accounts of these historical developments and analyses, extracts detailing developments in Ngā Manu Kōrero, presenting these as a continuing motif of Māori education.

Mua Te Whāia, Muri Te Taea: Using the Past to Describe the Present (Te Taura Whiri o te Reo Māori,[27] 1990)

Contemporary analyses of education in Aotearoa New Zealand commonly take as their starting point the formal schooling system that developed after the signing of the Treaty of Waitangi, in 1840. In doing so they ignore, render invisible, and dismiss the vibrant Māori education system that served Māori well before colonization. Prior to colonial contact, educational provision was a system established *by* iwi[28] Māori (iwi-centric curricula in design and intent delivered in iwi-Māori language), *for* iwi Māori (through pedagogical practice grounded in Māori epistemic and ontological views of the world), *with* iwi Māori (in a context where being Māori was the unqualified norm), *in* iwi Māori homelands (in tribal contexts). Within these contexts and practices, students were ecologically grounded in the broader metaphysical and cosmological narratives of Te Ao Māori[29] that were infused with narratives of lore and tikanga a iwi.[30] In short, in similar ways to other educational systems, Māori education provided experiences grounded in ontological, epistemological, axiological positions that affirmed for Māori their "cultural blueprint" and their normality (Jackson, 1998).

With the introduction of missionaries and their interest in the conversion of "natives" to Christianity, there was a consequent shift in emphasis (albeit ad hoc across iwi groups) that saw iwi education being imparted *by* missionaries (offering a Christian-centric curricula used explicitly to undermine and repress iwi-centric knowledge as requisite to embracing the new faith), *for* colonial objectives (through pedagogical practices grounded in colonial epistemic and ontological views of the world), *with* iwi (in a context where difference was forged at an ideological level but with peers and whānau who had significant whakapapa links to each other), *in* iwi homelands (in the tribal homelands that by and large remained within their control).

In the years following the arrival in 1814 of British missionaries in Aotearoa, developing Māori literacy became a focus of their educational efforts, with the earliest material produced and printed by Kendall (1815) in Aotearoa, New Zealand, being solely in te reo Māori. Despite the early missionaries' reports of Māori cosmology, spirituality, and epistemology as broad and sophisticated (Smith, L., 1999, 172) and the fact that by 1840 "a large proportion of the Māori population could read and write in their own language" (Beaglehole, 1970, 24), Māori people were represented by the colonial state as in need of "saving" from savagery and immorality.

It is worth exploring this area in some more detail to illustrate the dual dynamics of, on one hand, Pākeha colonizing aims of civilizing and assimilating Māori through literacy education, and on the other, Māori enthusiastic adoption and co-option of literacy for enhancing their own knowledge bases, economic aspirations, and political skills. One strategy of the colonial state was to "harness the power of print in its project to pacify, educate and modernize Māori," as Ballantyne (2006, 20) put it. This "mission" was articulated in an early newspaper's statement that it was selecting "materials best calculated to elevate and enlighten the native understanding and to render the Māori a fit and civilized associate of his English fellow subject" (Ballantyne, 2006, 20). Thus, while literacy was important, creating a population literate solely in Māori provided for Pākeha a way of controlling the flow of information, through Pākeha retaining the power to decide or prioritize what would or would not be translated.

Māori retained their own "way of doing things" though. The responses of one group of Māori to published editions of one of the Māori newspapers, Te Karere o Nui Tireni, were described by William Brodie (1845, 109–110) as follows:

> One native of a party is generally selected to read the news aloud. When he takes his seat upon the ground, a circle is then formed, and after the reader has promulgated the contents, the different natives, according to their rank, stand up and argue the different points contained; which being done, they retire home and answer the different letters by writing to the editor.

Ballantyne's comments on this are salutary for the purposes of our discussion of Māori education:

> With its emphasis on reading aloud in a communal setting and the structured verbal and written responses to these texts, Brodie's sketch suggests some of the ways the norms of an oral culture moulded responses to the printed word, reminding us that literacy and the newspapers did not

displace older knowledge traditions, but rather were interwoven with those oratory and political discussion traditions. (Ballantyne, 2006, 21)

While Māori interest in European education was based on expanding their knowledge bases, a common lack of interest by mainstream educational scholars in traditional Māori educational practices echoes, sadly, the persistence of assimilatory processes of educational policy, regulation and provision for Māori students after the failure by the Crown to keep the promises enshrined in the 1840 Treaty of Waitangi. The ensuing large-scale land purchases and confiscations had disastrous consequences for Māori people for many decades. Iwi suffered severe economic losses, with social and health effects that continue to exact a toll to date, and disproportionate representation in negative social indices. Further, in disconnecting different whānau, hapū, and iwi from their communally owned land, Māori people were disconnected from their ways of being and seeing the world, spiritually, politically, socially, and economically. They had to struggle not only to achieve "official" status as the tangata whenua in their own lands, but also to ensure the survival of te reo[31] and tikanga[32] Māori—their language and culture (Smith, L., 1989; Jenkins and Ka'ai, 1994; Walker, 1996).

Despite evidence of the vigor of Māori intellectual life and other evidence of the enthusiastic engagement of not only Māori leaders but also taurekareka[33] in reading and writing (Brodie, 1845), political debates on the education of Māori children were kept quite separate from the debates on the education of European settlers' children (Harker, 1985, 61). The underpinning assimilatory purposes of the policies developed by the colonial state are clearly echoed in early parliamentary debates about education. Accounts of that period are salutary for understanding current issues around Māori education—and for understanding the symbolic, cultural, and educational significances of an event such as Ngā Manu Kōrero speech competitions. Let us then expand now on what we see as some other significant historical events in Māori/Pākeha relations and education in our nation.

In 1849 Governor Grey subsidized with government funding the missionary boarding schools that had been established for Māori children, because these schools were useful in isolating Māori children from the "demoralizing influences of their villages" (Grey, cited in Beaglehole, 1970, 28). Then, during the later 1867 parliamentary debates around replacing the mission schools with day schools in Māori villages, Hall declaimed that the government's aim was to achieve the "civilisation of the remnant of a noble race" (Parliamentary Debates, 1867, 866, cited in Harker, 1985, 61). Carlton's view in this debate was that the colony could not afford the military exercise to "exterminate the natives" and, as Harker put it, "he plumped for 'civilising' them" (1985, 62). Under the resulting Native Schools Act 1867, Māori hapū were required to provide "one half of the costs of construction, one quarter of the teachers' salary" and enough land for the school in their area (Beaglehole, 1970, 29).

Throughout these years, the aim of the colonizing policy makers was to assimilate Māori, "Europeanize" them; and to facilitate this process, instruction was to be in English. While bilingual Māori–English texts were supplied to the first of the state-controlled Native schools, an inspector reported that "no Māori is allowed to be spoken in the school"—the use of the texts was tolerated "only to facilitate the learning of English" (Simon and Smith, 2001, 162). There is evidence that some te reo Māori continued to be used in the Native Schools though, both by teachers (who were expected to have a good understanding of the language themselves) and by children in the playgrounds (Simon and Smith, 2001, 163).

During the late 1880s, however, after the land wars and Māori land confiscations, disaffected Māori began withdrawing from state education. This concerned Māori leaders, whose focus for their people was on economic and political survival in a Pākeha-dominated and English-speaking society. Being fluent in English was an obviously important strategy, so they petitioned for an amendment to the Native Schools Act, "which would require the teachers of a Native School to be ignorant of the Māori language and not permit (it) to be spoken at the school" (Simon and Smith, 2001, 164–165). As Simon and Smith have noted, "Te reo was in a position of strength at this stage...it is unlikely that Māori ever conceived that its survival would come under threat as a consequence" (Simon and Smith, 2001, 165). Furthermore, the desire to focus on English fluency for Māori was in keeping with Māori desires to have access to a broader range of literature than was initially made available to Māori. In Barrington's opinion, the agreement of some Māori parliamentary members with the "Europeanizing" policies as potentially benefiting their people may have contributed to the increased rolls of the "native schools"—from 1,625 in 1880 to 10,403 by 1939 (Beaglehole, 1970, 37).

During the years up to the late 1930s, however, educational provision for Māori students in these schools was also shaped significantly by Pākeha political agendas for producing a pool of semiskilled and unskilled labor. The objective "seemed to be the development of a rural-based peasantry," like those in England and France (Harker, 1985, 63). Certainly, the curriculum provided for Māori students was aimed at leading the "lad to be a good farmer and the Māori girl to be a good farmer's wife" (Ramsay, 1972, 68). In the 1940s, by which time many Māori were moving to the towns, the new curriculum for Māori schools (reflecting persisting and wrong denigrating assumptions about Māori students' intellectual capabilities) focused on developing practical skills of building, plumbing and so on for boys, and cooking and sewing for girls. The effect was social containment and a limiting of job opportunities for young Māori people. Nevertheless, provisions were made for a system of scholarship to enable "bright" Māori children to attend denominational boarding schools with the idea that an educated Māori elite would then return to their papa kāinga[34] and spread the gospel of assimilation—thus helping to further Pākeha interests (AJHR, 1881, E-7, 1–11 cited in Simon, 1994). There were some exceptions, as in the case of the principal at Te Aute Māori Boys Boarding School, John Thornton, who coached promising students to compete and matriculate. This led to the first wave of Māori university graduates. Thornton faced the threat of reduced funding, however, if he continued to deviate from the prescribed curriculum (Barrington, 1988, 1992; Simon, 1994).

Although much was lost to Māori during these years, especially in eroding of te reo, Māori resistance ensured that neither te reo nor tikanga Māori were lost. There was also a growing recognition by politicians that "although assimilation worked in favor of Pākeha dominance, it did not serve to conceal social contradictions" (Simon, 1986, 12). Consequently, arguments for a shift in policy direction gathered momentum, culminating in the release in 1961 of the Report on the Department of Māori Affairs (Hunn, 1960). This report officially rejected assimilation and redefined Māori–Pākeha relations in terms of integration, which was "to combine, (not fuse) the Māori and Pākeha elements to form one nation wherein Māori culture remains distinct" (Hunn, 1960, 16). On first reading, this appears to sit within an (alter)native mainstream paradigm. However, the report rather contradictorily stated that there might need to be explanations of the advantages of integrating, given by Pākeha to Māori who "live a backward life in primitive conditions (or who)... resent the pressure brought to bear on them to conform to what they regard as the Pākeha way of life" if such Māori were to "fall into line." Indeed, as Harker and McConnochie (1987, 61) commented, this was a "strange kind of integration." In a later government publication, integration was further defined as:

> a dynamic process by which Māori and Pākeha are being drawn together, in the physical sense of the mingling of the two populations as well as in the mental and cultural sense, where differences are gradually diminishing (Booth and Hunn, 1962, 2).

It is clear from this statement that assimilation of Māori (as in our alternative paradigm) was still on the political agenda.

The Hunn Report did make an important contribution in that it clearly identified the numbers of school leavers with no educational qualifications.

It also advocated "positive discrimination" to redress the Māori disadvantage. However, this analysis was based on a cultural deficit model that saw Māori as the problem and overlooked the fact of prejudice on the part of those who defined the deficit. In other words, the "deprived" had to come up to Pākeha standards. The policy thus became

> ...ideological in the hands of teachers and administrators, and served instead the interests of middle-class Pākeha...When used to explain the under-achievement of Māori pupils, it serves to conceal the extent to which the education system is structured round Pākeha interests, values and goals (Simon, 1986, 8–9).

Ngā Manu Kōrero/Māori education:
The Push and the Parry, Subverting and Co-opting Space

Bringing Māori "up" to Pākeha standards was also part of the original rationale for the state's introduction of Ngā Manu Kōrero, the Māori speech competitions. An account of the competitions' ensuing development over the following decades provides us with a potent illustration of Māori aspirations and demonstrates one aspect of a Māori willingness to simultaneously engage on multiple fronts in their ongoing struggles against hegemony and the Pākeha education system's assimilatory effects. In the following parts of the chapter, therefore, we juxtapose (slightly edited) extracts from Ngā Manu Kōrero's official Web site[35] (which details the evolution of Ngā Manu Kōrero) with some details of contemporaneous state policies and Māori initiatives related to Māori education.

It is salutary that the 1960 Hunn Report, first celebrated for its shift in focus from hard-line assimilationist policies to "softer" integration, continued to endorse prior colonial beliefs about Māori language as a "relic of ancient Māori life." Unsurprisingly, the introduction of Ngā Manu Kōrero in 1965 aimed to encourage a greater command and fluency of spoken English amongst secondary Māori students. First known as the Korimako Speech Contest, the competition was sponsored by the then governor-general, Sir Bernard Fergusson (Lord Ballantrae). The competition, held annually, was initially open to all Māori students in any secondary school across the nation. Where there was more than one competitor from any school, school competitions were held. The winners of these school competitions then competed regionally to represent their cluster of schools at the national competitions. Māori embraced the contest (as they had embraced Māori Boarding Schools) as offering further educational opportunities for their children. That English was seen by Māori individuals, hapū and iwi, as just *one* component of the competitions rather than their totality is illustrated in the trophies that were created and presented by different groups. Although this first competition was about celebrating the acquisition of English, the carved trophy awarded to the winner remains a significant symbol for Māori, as Māori. It is a carved warrior in traditional attire, holding, in a stance of readiness, a taiaha.[36] This trophy can be read in multiple ways—as Māori pride in things Māori, as a symbol of courage or leadership, resistance, defense perhaps, or determination. The runner up receives a waka huia,[37] in which chiefs or family groups customarily stored highly prized huia feathers and other taonga.[38]

Korimako (Senior English)(1965)

The Jock McEwen taonga is awarded to the runner-up in the Senior English contest. This taonga, a waka huia, was carved and presented in 1965 by Mr. Jock McEwen, a former Secretary of Māori Affairs.

The competition remained as an English-only speech-making contest for the next twelve years. Throughout these years, however, beyond the competition there was a growing concern over access and retention rates of Māori in education. The disparity in educational achievement between Māori and Pākeha did not improve. In 1974, the Māori Education Foundation Report noted that 73 percent of Māori school leavers had no educational qualifications, compared with 34 percent of non-Māori. Deficit theories and theories of deprivation, which saw Māori culture and values being the root cause for individual Māori underachievement, prevailed as explanations for the disparities. As Linda Tuhiwai Smith later commented:

> ...by concentrating on educating or influencing the individual, there is a tendency to localise the issues and solutions...Organisations can distance themselves from an individual response and can continue to function in ways that still seek to acculturate or destroy...Pākeha people are able to define what is valid, worthy, useful and right. (Smith, L., 1986, 5–6)

In Bourdieu's terms, educational institutions remained structured to favor those already possessing the dominant cultural capital.

In contrast to accounts that located Māori failure in being Māori, however, Māori continued to assert the centrality of Mātauranga[39] Māori to their survival as a people. This was no longer primarily about surviving physiologically (as had been the challenge at the beginning of the twentieth century) but about being able to "survive as Māori, to retain a Māori identity, while still being able to participate fully in society, and in the communities of the world" (Durie, M., 1997, 1). On the back of Benton's prediction (1981) that the Māori language was on the brink of extinction, new forms of alternative/(alter)native education were born. Action and reaction from both Māori and Pākeha marked a new era in Māori Education. As Māori continued to struggle to achieve positive educational provision for Māori outside the system, state policy makers took a new approach to thinking about their role. Consequently, Taha Māori, a state educational policy and initiative, was introduced in 1975.

Originally endorsed by Māori people, Taha Māori aimed to infuse a Māori dimension into every aspect of school curricula, policies, and practices. "Māori people themselves created the programs and resources, trialed these programs and helped educate colleagues to carry them out" (Smith, L., 1986, 7). Sadly though, as Smith pointed out, "in the long term it has served only to perpetuate the structure and practices of the educational bureaucracy" (Smith, L., 1986, 7) with Māori dimensions largely "divorced from their cultural context and incorporated in terms of the pedagogy and evaluation systems of the dominant group" (Harker, 1985, 69). Graeme Smith concluded, "Taha Māori has only limited relevance to the concerns and aspirations expressed by Māori people" (Smith, G. 1986, 17). Moreover, because it was directed at *all* children, in the hands of mainly Pākeha teachers, Taha Māori defaulted to educating Pākeha students in biculturalism rather than strengthening a Māori sense of identity and redressing Māori cultural loss. If we think about this in relation to the depiction in figure 8.1 of what constitutes Māori education within the mainstream, Pākeha teachers teaching Pākeha students about things Māori constituted the primary focus. Taha Māori drew, however, from Māori funding, and these teachers were doing little to advance Māori education for Māori students.

Given the preceding events and dynamics, it is not coincidental then that in 1977 Ngā Manu Kōrero was expanded to include a section for the best speaker in Te Reo Māori (Forms 5–7).

Pei Te Hurinui Jones (Senior Māori) (1977)

This section, open to all New Zealand students, regardless of ethnicity, focuses on encouraging all secondary students to speak both English and Māori languages. The taonga awarded to the student who gains the highest marks in both the prepared and impromptu sections of the Senior Māori contest was presented by the Māori Education Foundation to commemorate the life work of the late Dr Pei Te Hurinui Jones, a distinguished Waikato elder and scholar who died in 1976. Dr Jones was a renowned orator and prolific writer. The extent of his knowledge of Māori traditional lore is exemplified in his revision of the Ngā Moteatea[40] series. The University of Waikato awarded him the degree of Doctor of Literature for his contribution to literature.

The following decade seemed to herald a more substantive change in thinking among state policy makers. In 1980, in *He Huarahi*, the report of National Advisory Committee on Māori Education, it was stated that cultural diversity should be *valued*.

> This report re-emphasises altering the attitudes and improving the knowledge of the Pākeha majority, working on the principle that if these aims are achieved, many of the other changes needed will tend to follow. (NACME, 1980, 12)

In Simon's later judgment, however, the idea in the report of respecting cultural differences was still used normatively: it referred "to deviations from the norm of the 'real' culture—Pākeha culture (which were) still seen as a problem to be overcome" (Simon 1986, 14). Thus, in this report Māori education continued to be located primarily within an *alter-native (alter-the-native)* paradigm, wherein the valuing of English as the primary medium for learning is paramount.

In the same timeframe, Ngā Manu Kōrero introduced a new Junior English section (Forms 3–4).

The Sir Turi Carroll taonga (1980)

The Trophy, a greenstone taonga presented by Ngāti Kahungunu ki Te Wairoa[41] *is awarded to the Māori junior student who gains the highest marks in a prepared English speech. It commemorates the late Sir Turi Carroll OBE, revered Kahungunu leader who stood firmly in both Māori and European worlds. He was the inaugural Chairman of the New Zealand Māori Council.*

While embracing each new addition to the competition as a marker of history, Māori resistance to emphasis on English *(alter-native)* forms of education continued. Ngā Kohangā Reo[42] early childhood education centers were established. Based on the concept of whānau and with huge voluntary input from many Māori women, there was rapid growth of these centers—by 1991 there were 630 centers with 10,451 Māori children enrolled (Davies and Nicholl, 1993, 20). Issues of cultural and language survival and whānau development were central for those who began the kōhanga movement: as Johnston (1993, 3) later noted, "Kōhanga had to be controlled, defined and legitimated by Māori for Māori." Sadly, though, the parent "grassroots" decision-making and control that prevailed for the first eight years was undermined when, struggling under severe financial hardship outside of the state system, Māori agreed in 1990 to being funded by the state. Consequently, kōhanga became subject to state policy and accountability requirements, and in Johnston's view its kaupapa risked becoming corrupted (Johnston, 1993, 14).

The tensions between Māori historical aspirations for (alter)native education (where Māori would have their place within the mainstream alongside their treaty counterparts) and the ongoing push by the state for alter-native assimilatory policies and practices are evident in the above account of the kōhanga movement. These tensions fed Māori resolve to have alternative parallel programs. In another initiative to bolster the survival and growth of te reo, in 1983 a Junior Māori section (Form 3–4) was added to the Manu Kōrero speech competitions.

> **The Rāwhiti Īhaka taonga (1983)**
>
> *Awarded to the junior student who gains the highest marks in a prepared speech in Māori. In 1983, senior pupils of St Stephen's (a Māori boys boarding school), led by their Headmaster, Scottie McPherson, presented the Trophy as a tribute to a teacher and mentor, commemorating his life and his oratory skills in both Māori and English and his skills in teaching mathematics and science.*

In September 1983, however, the first Core Curriculum Review Working Party was convened by the minister of education with no Māori representatives (Walker, 1985, 78). This omission was perhaps the last straw for those who had been working for years to advance the education of Māori students. Frustration over the state's failure to provide an equitable (alter)native education for Māori within the compulsory sector came to a head. A large number of Māori came together at the Māori Educational Development Conference held at Tūrangawaewae in March 1984 to share their concerns and plan a different way forward. Maiiki Marks, a Māori language teacher, was one speaker who described the frustrations of her job and feeling of being a mourner at the tangihana[43] of her own culture (Marks, 1984, 6). Marks was not alone in her view—at the end of the conference, the participants recorded that:

> The existing system of education is failing Māori people and modifications have not helped the situation, nor will they. Therefore we call for withdrawal and the establishment of alternative schooling modelled on the principles underlying Kōhanga Reo.

This was a call for Māori *alternative* schools as *separate institutions parallel to state schools* (see Durie, 1993, cited in Court, 2005). It resulted in the establishment of the Kura Kaupapa Māori movement, with the first kura, offering primary (year 1–6) education in te reo me ono tikanga

Māori, being opened at Hoani Waititi Marae in 1985. In the same year separate awards celebrating the linguistic skills of both males and females were donated to the Ngā Manu Kōrero competitions, adding to the growing number of trophies commemorating the lives of Māori committed to education.

> ### Ngā Kete O Te Mātauranga[44] (1985)
>
> *Awarded to the female who gains the highest marks in the Pei Te Hurinui Jones (Senior Māori) section of the competition. The Trophy, a korowai[45] and carved base, was presented in 1984 by the people of Te Tai Tokerau.*
>
> ### The Riki Ellison taonga (1985)
>
> *Awarded to the male who gains the highest marks in the Pei Te Hurinui (Senior Māori) section of the contest. Presented by the family of Riki Te Mairaki Taiaroa Ellison as a memorial to their pōua,[46] to acknowledge his influence in revitalising Te Reo Māori in Te Waipounamu.[47] The trophy is a kauri[48] toki[49] wedged in a tōtara[50] base with a pounamu[51] inlay.*

The critique of and conscious withdrawal from compulsory education by many Māori trained teachers and students impacted on the Pākeha mainstream in a number of ways. When the second Curriculum Review Committee was convened, Māori were included. This committee undertook wide consultations with public groups and individuals throughout 1985, canvassing opinions on a range of topics, including issues relating to Māori and to gender in education. As the committee considered the huge number of submissions that came in and prepared their report during 1986, it became clear that education was facing a legitimation crisis in the area of Māori and the curriculum. Some other significant events that were occurring during this time highlighted the issue of language and undoubtedly influenced both those who were making submissions and the Curriculum Review Committee members' thinking. In 1985 a Māori claim before the Waitangi Tribunal asserted that language was a cultural treasure and as such protected under the provisions of the Treaty of Waitangi. Following the Tribunal's affirming this claim, Broadcasting New Zealand established a Māori radio board in 1986, and the Māori Language Act 1987 made Māori an official language. The Curriculum Review final report, presented to the Minister of Education in late 1986 and published in early 1987, unsurprisingly reflected a diversity of views on the topic of Māori and Pākeha and the Curriculum. However, it promised that the principles of the Treaty of Waitangi on Māori language and culture would be honored in the curriculum. And, more radically, it shone a spotlight into a murky part of mainstream education that had long escaped official attention: Principle 3 stated that "The curriculum shall be *non-racist*" (Department of Education, 1987). Kath Irwin, a Māori academic, later responded, "This statement is a vital inclusion... Were it stated in any 'weaker' form, which omitted the word racism, the intention to plan seriously for the future education of all New Zealanders...... would be lost" (Irwin, 1988, 49).

This shift in mainstream curriculum policy toward (alter)native programs did not, however, check the growth of Kura Kaupapa Māori as a *parallel alternative* for Māori. In spite of lack of funding, increasing numbers of Māori parents and educators with few resources took up the challenge of establishing and participating in schools that enabled Māori decision-making on their own terms, about all matters of concern to their children's education. Like Kōhanga before them, however, they eventually had to seek state funding. Although this request was granted, funding was limited to supporting only five new kura a year, slowing the growth of this movement.

Arguably, these funding decisions were part of the downstream effects of the new right restructuring of the welfare state instigated earlier in 1984, a "revolution" (Kelsey, 1995) that aimed to increase the efficiency, effectiveness, and accountability of public sector provision (see Boston et al., 1996, for a detailed discussion). In 1988, the consequent restructuring of the educational administration system (Department of Education, 1988) turned the focus of educational policy makers and practitioners onto the development and implementation of standardized market managerial systems of surveillance and control. This had the effect of marginalizing for nearly a decade social justice issues such as those we have been discussing (see Court, 2003; 2004 for more detailed discussions).

While the critique of the 1989 Education Act gained momentum, Ngā Manu Kōrero introduced its first taonga that was not specifically linked to a competitor and the first trophy to have an inscription and explanation solely in Māori. As made clear in the Māori inscription below, the trophy was gifted to honor the integrity of the competition itself. The trophy, a carved canoe, was specifically named as the symbol of the mauri[52] or spirit of the competition, as a symbol of pan-tribal unity. Those gifting the canoe intentionally avoided reference to region or tribal group, wanting instead to remember all ancestors who had passed before. The canoe, as the vessel that all Māori tribes used to traverse the Pacific, is a powerful symbol of life, of journeys travelled, and those yet to be travelled. The hoe or carved paddle was added nine years later. These trophies are passed from one national host to the other with significant ceremony at the completion of national competitions.

> **Te Mauri[53] o Manu Kōrero Trophy (1989)**
>
> *Presented as a material symbol of the hidden principle protecting the vitality and mana of Ngā Manu Kōrero. The inscription on the taonga reads: He waka tohu aroha na te Wairarapa. I te kōrero a te kaumatua nei, kia kore ai e tumekemeke ngā iwi. No reira, ko tēnei tāonga he tāonga nā ngā mātāwaka o runga o Aotearoa.[54]*
>
> *In 1998, a carved hoe[55] was added to the Trophy, presented by Tai Rāwhiti in memory of Rose and Joss. Kia noho tahi ai te waka me te hoe hei kawe i te mauri mō Ngā Manu Kōrero.[56]*

Ngā Manu Kōrero continued, increasingly drawing on Māori for support in regions throughout Aotearoa, New Zealand. New trophies demonstrated both regional and pan-tribal affiliations, as illustrated in the first case below. While each of these trophies are awarded at the national competitions, they are taken by the winners back to their own regions until the commencement of the next series of regional competitions where they are once again displayed.

> **Kiri Moerangi Mangu taonga (1991)**
>
> *Awarded to the student who gains the highest marks in the Senior English impromptu section. The trophy was presented in 1991 by Southland Girls High School as a memorial to Kiri Moerangi Mangu, of Ngāti Porou, who won the Regional Otago/Southland Korimako Speech Contest in 1990. Her goal was to participate in the National Ngā Manu Kōrero finals but she passed away at the end of Term 2. The trophy is a framed photograph of kotuku.[57]*
>
> **The George, Te Kēpa and Hamiora Stirling taonga (1991)**
>
> *Awarded to the student who gains the highest marks in the Senior Māori impromptu section. The Trophy was presented by Te Keepa and Pani Stirling as a memorial to their sons, who dedicated their lives to Te Reo Māori. It is a carved waka huia.*

The continued Māori push for parallel alternative education resulted in the establishment of three Wānanga[58] between 1993 and 1997. Of the three, two are tribally based and are accredited to confer degrees to doctoral level, pushing traditional universities to parry this Māori drive into tertiary education. In 1994 our own College of Education (prior to merging with Massey University) introduced Te Tohu Pokairua, an undergraduate teaching-diploma taught wholly in Māori. In this same year, Manu Kōrero added a trophy to celebrate the bilingual student who could compete in both languages.

> **E Tipu e Rea Trophy (1994)**
>
> *Commemorating the centenary of the capping of the first Māori graduate at a New Zealand University, Sir Apirana Ngata. This taonga is awarded to the student who gains the highest aggregate in both Māori and English languages.*

Then in 1998 a special event occurred at Ngā Manu Kōrero.

> **The Māori Education Trust (1998)**
>
> *A special presentation was made to all the national finalists in the Ngā Manu Kōrero, and their schools, to commemorate Sir John Mokonuiarangi Bennett, QSO, Chairman of the Māori Education Foundation/Trust for 25 years and the competition's inaugural patron.*

And in 2000 at our College of Education, a full immersion Māori Bachelor of Education program, Te Aho Tatairangi, was established. The College has since celebrated a number of graduates from this program, who have gone on to contribute to alternative Kura Kaupapa Māori and (alter)native programs. Notably, in 2004 the first Massey University student who had been taught solely in Māori throughout her education graduated. Her educational experiences included alternative (Kōhanga Reo and Kura Kaupapa Māori) and (alter)native spaces (Massey University College of Education). She was also a previous recipient of the E Tipu e Rea Trophy.

He Waka Kōtuia Kāhore E Tukutuku Ngā Mimira
A Canoe That Is Interlaced Will Not Become Separated at the Bow

Today, we may look back on the introduction of state-provided education for Māori students with a critical eye, judging the early assimilatory and later integrationist aims as colonizing practices. In continuing to speak of "Māori education" only in this way, however, we risk missing the fissures and frayed edges that mark the diverse creativities and chafing of people, pedagogies, and cultural practices. We could only begin to sketch some of these complexities in this chapter, as we tried to view Māori struggles in education occurring in relational spaces of dynamic cultures in constant interplay.

Although in this chapter we have been interpreting these struggles through a Māori view of using history to describe the present, we suggest that none of the kuia and koroua we have represented here saw themselves as "the done to lot." Instead they saw themselves as active agents working to shape histories in the contexts (as difficult as these were) in which they found themselves. Simon (1994) noted that from the point of contact both Māori and the Crown displayed intense interest in educational provision for Māori. What Simon and Jenkins (1991, cited in Simon 1994) also pointed out, however, is that while both saw value in it, expectations about what that schooling was meant to provide varied markedly.

> On the surface the intentions of the government in regard to schooling appear to be similar to those of the Māori. Yet when we look more closely we see that they are essentially different. Māori embraced schooling as a means to maintain their sovereignty and enhance their life chances. The government, on the other hand, sought control over Māori and their resources through schooling. Māori wanted to *extend* their existing body of knowledge. The government, with its assimilation policy, intended to *replace* Māori culture with that of the European. (Simon, 1994, 58–59)

It is unsurprising then, that Māori engagement in state education has oscillated between ardent participation at times to conscious withdrawal at others.

In our reflections on education for Māori, we have also been trying to indicate that, contrary to an implied state notion of one schooling provision for Māori, since the beginning a defining feature of Māori educational initiatives by Māori in Aotearoa New Zealand, is that these are not focused on a singular or generic approach. Rather, in keeping with a Māori worldview, which Jackson (2007) maintains is grounded within notions of whakapapa, it has been accepted and expected that there is more than one way of doing things. Jackson argues that the "very notion of our whakapapa implies generations of different stories layered on top of one another" (172). Telling stories (as students do in Ngā Manu Kōrero) was always a journey to the point of enlightenment that was known as the explanation or whakamarama.[59]

As the Te Māuri o Manu Kōrero trophy suggests, Ngā Manu Kōrero competitions are seen by Māori as an embodied ceremony that in its essence, or life force, echoes both the historical and contemporary landscapes of Māori–Pākeha relations and practices in education. As we have shown, historically initiated by the state, Ngā Manu Kōrero occurred within a Māori context between Māori students. While the Māori students have always represented distinctly different hapū and iwi, they also come now from very different kura/educational programs. As such, this competition is grounded in and reflects the internal complexities of educational provision for Māori, with Māori, and about things Māori. Recently, while judging Ngā Manu Kōrero competitions, it occurred to Hine that (with the exception of home schooling and native schools that finally closed in 1969) embodied in the youth before her were the multiple and variously textured threads that constitute Māori education as modern forms of indigenous education, born out of political struggles that continue to evolve.

There, in that hallowed school hall,

standing, sitting or shuffling forward, are students drawn from

*Māori boarding schools (missionary schools predating the Treaty of Waitangi),
who sit alongside those that represent the biggest political win of all –
Kura Kaupapa Māori (immersion Māori medium schools) - both tribal and
pan-tribal kura.*
*These students, one group from the oldest and the other from the newest form
of educational provision (since colonial contact)
nudge their peers in
general-stream state schools.*

*Some of these students are in linguistic and cultural enclaves –
rumaki or total immersion programmes in larger schools
bilingual programmes
others hail from schools where students take reo Māori as a foreign language
much as they would French or Japanese
still others, taking Māori as a single subject, are required to enrol in
correspondence school
and still more who may not have the opportunity or inclination
to take Māori language at all…*

*They come from
State,
co-ed,
single sex,
Church based and
private schools.
All decile ranked, marked and rated 1–10 -
all here ready to compete, to debate, to provide their social
critique of topical issues confronting
te ao Māori in contemporary society.*

In these ways Ngā Manu Kōrero is emblematic not only of the rich variety of Māori educational experience, initiatives and programs, but also of what might seem to some to be a paradoxical unity of Māori peoples within their diverse realities. Kuia and koroua would have recognized and accepted as "normal" this paradox of diversity in unity. Traditionally, heated debates on all manner of social issues occurred internally within and between hapū and these debates were not only relished but revered.

The debates were mediated, however, by a unifying impulse within and across tribal groups when external forces worked to erode or eradicate their existence (Carlton, 1867, cited in Simon, 1994). An important example of the latter occurred when colonizing forces in this country provoked the emergence of the Kīngitanga movement in 1857. Chiefs from diverse iwi came together by the waters of Lake Taupo to debate the future for their peoples. They confronted the dilemma of whether to remain distinctly different peoples (iwi) or to unify as Māori against the external colonizing threat. Subsequently, in 1858, after Potatau Te Wherowhero was elected as the first Māori King of a confederation of iwi, he asserted to the people:

> Kotahi te kōhao o te ngira, E kuhuna ai te miro mā, te miro pango, te miro whero. A muri I a au Kia mau ki te aroha, kia mau ki te ture, kia mau ki te whakapono.
> There is but one eye of the needle through which the white the black and the red threads must pass. After I am gone hold fast to love, to the law and to all we believe in. (King Potatau Te Wherowhero, 1858)

While we know that he saw the threads as different peoples, Māori and Pākeha, coming together as they passed through the eye of the needle, it is not clear how he envisaged the latter. Our discussions in this chapter point to a reading of Potatau's whakataukī (proverb) as having three components. First, Māori and Pākeha are multiple peoples. Second, the common "eye of needle" all pass through is education; and third, within that portal there is an entwining of Māori-centered/different Māori programmes and state education programs, as well as an interlacing of Māori peoples from different hapū and iwi with Pākeha. Potatau was presenting each thread passing through the needle as of equal weight, held together in equilibrium. This view means that for each group of peoples, the passage through the education "eye" should not cost them their identity and ability to be self-determining peoples. This is especially important for Māori, whose identity as tangata whenua and as members of distinct hapū and iwi has for so long been under threat of eradication. But equally important, retaining a Māori identity should not be constructed as prohibitive to engaging with wider bodies of knowledge. It is clear that historically Māori did not conceptualize themselves as "alternative." Rather they saw themselves as within the state education system because as Treaty partners they were clearly mainstream—equal to their Treaty counterparts. This was not to bleach out their tribal distinctiveness or differences within their unity, but to retain a distinct sense of being Māori in the mainstream.

Although our discussions and conceptualizing of historical and current Māori education events as threads of *alternative/alter-native/(alter) native* forms of education have focused on implications for Māori peoples, these events and struggles have not just variously shaped Māori. According to Banks (1997), mainstream curriculum developed within the different kind of fourth-world *alter-native* framework we have critiqued disadvantages not only minority group students, dominant group students can be disadvantaged also, in at least four ways:

- it tends to reinforce their false sense of superiority;
- it provides them with a misleading conception of their relationship with other racial and ethnic groups;
- it denies them the opportunity to benefit from the knowledge, perspectives, and frames of reference that can be gained from studying and experiencing other cultures and groups; and
- it denies dominant students the opportunity to view their culture from the perspectives of other cultures and groups.

It is clear that within alter-native approaches in Aotearoa, New Zealand, Pākeha students have received skewed educational experiences that miseducate (Waitere-Ang and Adams, 2005). We maintain that adopting instead a bicultural frame of reference can provide opportunities for both Pākeha and Māori to recognize the socially constructed nature of education (and the ways we see the world) as a precursor to differently valuing Māori as *(alter)native* mainstream.

We maintain also that locating Māori education in its rightful place as an *(alter)native* approach in the mainstream will benefit all in our country. Already we can see how some of the diversely textured strands of Māori initiatives woven into the borders have bled into the centre, impacting on mainstream education and educators. For example, within New Zealand universities in recent times, space has been opened up by Māori academics for the development of "decolonized" kaupapa Māori and te ao Māori research methodologies (Smith, L., 1999). Using these approaches that value Māori views and recommendations, increasing numbers of researchers are investigating and analyzing cross-cultural issues of concern in a range of educational sites. In our own institution, to give just a few examples, Māori academics and postgraduate students have investigated strategies for school cultural self-review and produced guidelines for change (Bevan-Brown, 2003), critiqued the Anglo/European leadership archive and described Māori women's experiences in leadership positions across the different programme types in the primary sector (Waitere-Ang, 1999), critiqued notions of Treaty partnership in state education policy (Graham, 2002) and in Pākeha teachers' varied understanding and limited implementation of the partnership principle in their everyday mainstream school practices (McLeod, 2002), and evaluated a mainstream school's initiation of a bilingual unit (Prosser, 2004). Pākeha academics and postgraduate students have investigated the experiences of Māori parents, board members, and teachers in a unique three-stranded school (Court, 2003; 2005); researched Māori Board of Trustee members' experiences in and recommendations to their mainstream schools (Turner, 2005); gathered parents' (the majority of whom were Māori) views of how a mainstream school could better provide a bicultural learning community in which all students were valued and successful (Dow, 2006); and researched the place of emotion in critical pedagogy using kaupapa Māori methodology in an iwi context (Williams, 2007). And together, we (Hine and Marian) have studied a bicultural model of coprincipalship (Waitere-Ang and Court, 2004).

Ngā Manu Kōrero competitions illustrate also some of the positive effects of Māori initiatives for change. While presently still taking place within Māori contexts and involving mainly Māori students and their supporters, Ngā Manu Kōrero competitions now include also some competitors and supporters who are Pākeha. These Pākeha students compete alongside their Māori peers in te reo section of the competition. As such, Ngā Manu Kōrero competitions provide a beacon, lighting a way forward to new dynamic bilingual spaces of education that embrace Māori education as an integral and equally respected part of the centre. If such spaces are to be founded equally in te ao Māori and Pākeha traditions, and be places where Māori, Pākeha, and new immigrant teachers and students can together test their views, share insights, learn, and grow, it will require all of us to leave some of our comfort zones. In our view, this would certainly change both "mainstream" and "Māori education" for the better.

Whiria te muka o te harakeke ki mua, nō te mea kei muri rātou e tatari ana.
Fling the strands of the flax rope forward to the future,
because behind they are waiting.

Notes

1. In this chapter we use a significant number of Māori words. This is an increasingly common practice in scholarship and current government policy. It recognizes the status of Māori as an official language in Aotearoa New Zealand. For international readers we provide footnoted translations for the first time a Māori word or phrase appears in the text.
2. Ngā Manu Kōrero, title of the competitions, literally means The Speaking/Talking Birds or Orators.
3. Whakakpapa—Genealogy.

4. te reo Māori—The Māori Language.
5. Tihei mauri ora—The sneeze/breath of life, a call to claim the right to speak.
6. Tūpuna—Ancestors.
7. Kuia—Respected female elder.
8. Koroua—Respected male elder.
9. Whaikōrero—Formal speeches with defined components.
10. Whakataukī—Proverbs.
11. Karanga—Ceremonial call provided by women.
12. Oriori—Lullaby, composed on the birth of a chiefly child about his/her ancestry and tribal history.
13. Pātere—Song of derision in response to slander, often with political focus.
14. Waiata—Songs.
15. Mātauranga—Knowledge.
16. Pākeha—Term used for non-Māori, generally of European descent
17. Team members in its development were: James Graham, Patricia Johnson (initial paper coordinator), Peti and Pani Kenrick, Jenny McLeod (current paper coordinator), Brian Paiwai, and Hine Waitere. For further comment on the papers, role, and function within the college, see Johnson and McLeod (2001).
18. Tangata whenua—People of the land, indigenous.
19. This term is used by Castells (1998), a Spanish sociologist, to explain a "lost" world or subpopulation subjected to social exclusion in global society. Different from the first-, second-, and third-world designations, which rank nation states according to their economic and political status, a Fourth-World label denotes nations or peoples without states. Defined this way Fourth-World encompasses presettler indigenous peoples whose economic status and oppressed condition arguably place them in an even more marginalized position in the sociopolitical hierarchy than other postcolonial peoples (Brotherston, 1992, cited in Ashcroft et al., 2000).
20. The Declaration of Independence was recognized by the British crown five years prior to the Treaty of Waitangi and gazetted in Britain. The Declaration recognized the sovereign status of Māori chiefs in Aotearoa New Zealand. This affected the way in which the Treaty process was to subsequently play out. For an elaborate discussion, see Cox (1993).
21. For a fuller discussion on the Treaty of Waitangi, see Claudia Orange (1987).
22. Kura Kaupapa Māori—Māori language immersion schools.
23. Rumaki—Immersion units within general-stream schools.
24. Pōwhiri—Formal ceremony of welcome.
25. Whānau—Family, which for Māori goes beyond the Western notions of the nuclear family.
26. For an explanation of Māori parallel systems as constituted on a bicultural continuum, see Durie (1993); and for a discussion of a Māori teacher's attempts to develop her classroom in this way in a unique school led by coprincipals, see Court (2005).
27. Te Taura Whiri o Te Reo Māori is the Māori Language Commission established as a Crown entity after the Māori Language Act 1987.
28. Iwi—Tribe.

29. Te Ao Māori—A Māori world view.
30. Tikanga a iwi—Tribal customary practices.
31. Te reo—The language.
32. Tikanga—Correct procedure, process or practice.
33. Taurekareka—Captive taken in war.
34. Papa kāinga—Original home.
35. http://www.Māorieducation.org.nz/mk/ accessed August 2007. See also Tinirau (2006).
36. Taiaha—Māori spear used to both thrust and parry.
37. Waka Huia—A carved treasure box.
38. Taonga—Treasures.
39. Mātauranga—Knowledge.
40. Ngā Moteatea—A lament or traditional chant.
41. Ngāti Kahungunu ki Wairoa—The name of an east coast tribal group. The particular subtribe indicated here is located in Wairoa.
42. Kōhanga Reo—Early childhood language nests.
43. Tangihanga—Funeral, rites for the dead.
44. Ngā Kete O Te Mātauranga—The baskets of knowledge.
45. Korowai—Cloak.
46. Pōua—Old person, grandfather.
47. Te Waipounamu—South Island.
48. Kauri—A native wood.
49. toki—Adze or axe.
50. tōtara—A native wood.
51. pounamu—Greenstone, jade or nephrite.
52. Mauri – essence or life force
53. Te Mauri o Manu Kōrero—The spirit or essence.
54. The inscription reads: This canoe is a symbol of love from Wairarapa as a pan-tribal treasure/symbol belonging to all the canoes of Aotearoa so that the awe of the tribes would not be lost.
55. Hoe—Paddle.
56. The canoe and the paddle reside together to carry the spirit of the competition forward.
57. Kotuku—White heron.
58. Wānanga—A place of higher learning, University.
59. Whakamarama—Illuminate.

Chapter 9

Starting with the Land: Toward Indigenous Thought in Canadian Education

Celia Haig-Brown and John Hodson

Standing on the earth with the smell of spring in the air, may we accept each other's right to live, to define, to think, and to speak.
—Eber Hampton (Chickasaw)[1]

The voices of these victims of empire, once predominantly silenced ... have been not only resisting colonization in thought and actions but also attempting to restore Indigenous knowledge and heritage. By harmonizing Indigenous knowledge with Eurocentric knowledge, they are attempting to heal their people, restore their inherent dignity and apply fundamental human rights to their communities. They are ready to imagine and unfold postcolonial orders and society.
—Marie Battiste (Mi'kmaq)[2]

Introduction

Let us begin by acknowledging that this chapter has been written on the traditional lands of the Mississauga people of the great Anishinaabe Nation, those of the Huron, the Neutral, and the Petun, and the lands of the Hotinonshó:ni, of the Six Nations located in what some now call the province of Ontario in Canada. We locate all of what we do in relation to the lands and the Aboriginal peoples who have lived with those lands. We bring that belief to bear on the thoughts we lay out for the reader in this chapter. Persisting questions guide us: What happens when Indigenous[3] ways of knowing and being in the world, exemplified in this recognition, come to bear on Eurocentric[4] forms of education and schooling? What are the possibilities for transformation of understandings, for shifts in world view, with deliberate efforts to interrupt classroom epistemological business as usual? In this chapter, using circlework teachings (see, for example, Graveline, 2003), we explore some responses to these questions, ever conscious of the fundamental significance of relationships to the investigation. Remembering, Resistance, Responsibility, and Regeneration name the patterns of thought focusing our discussion of Aboriginal education in Canada. Articulating these dimensions as sites of historical significance, accomplishment, potential, and possibility resonates with the work so many Aboriginal educators and their allies are doing in Canada at this time and provides ongoing inspiration for considering the place of Indigenous thought in all classrooms.

Starting with the Spirit of the Land

Education in Canada, broadly defined, begins, and always has begun, in relation to the land and Aboriginal peoples. Since time immemorial, Aboriginal peoples have integrated lifelong teachings and learning into the everyday worlds of their communities. Always reinforcing the relations of all beings to one another, the land becomes the first teacher, the primary relationship.[5] The land for traditional Aboriginal cultures in Canada is a complex being—a spiritual and material place from which all life springs.[6] As Makere Stewart-Harawira (2005) (Maori) says in her development of an Indigenous global ontology, "A central principle of indigenous peoples' relational ontologies and cosmologies is the inseparable nature of the relationship between the world of matter and the world of spirit." Aboriginal education in Canada, at its best, realizes and engages with this complex relationship.

In this context, "land" is recognized, in the sense of coming to know again, as so much more than a word. The land is physical: people walk on it, they literally put their feet on it, sometimes insulated by layers of concrete, pavement, flooring, and shoes, sometimes barefoot on bare ground. Everyday the land supports us in our journeys. The land is Spirit[7] and if we pay attention, the land speaks to us. It teaches us and all beings ways to live in good relation to one another. If we refuse to listen, the land speaks back with increasing clarity, sometimes returning to us the poisons we insist on feeding it. When we do pay attention, the resilient land heals herself and leads us to heal ourselves.[8] The land is the culture: Indigenous thought is built on such understandings. Canadians are only beginning to acknowledge that unlike the land portrayed in the lies that lured so many people to immigrate in good faith to these shores, this never has been an empty land.[9] Many of us are only beginning to ask that fundamental question, "Whose traditional land am I on?" For those who take the time to learn, there is always an Aboriginal nation engaged with the lands before any non-Aboriginal people arrived. Aboriginal people have named, lived in, and travelled through these spaces for thousands of years, since time immemorial, some say. And they continue to do so. The Spirit is in and of the land.

Geography Lesson

To do justice to Canada and the First Nations that make up this country in one small chapter is an impossible task. Canada is 3,854,085 sq. mi or 9,984,670 km. sq. It stretches from sunrise over the Atlantic to sunset in the Pacific, and north to the Arctic Ocean. Its southern door is bordered by the United States. It includes much of the Great Lakes, a long stretch of the Rocky Mountains, and several other new and old mountain ranges. It includes desert, muskeg, tundra, prairie, and rainforest. All of these places were known and named by the original peoples. Many of them have new names and some of them are reverting to the old ones.[10] All of these places knew and shaped the people in particular ways as they taught the people how to live with them.

Historian and member of the Métis Nation, Olive Dickason, reminds us, rather than seeing England and France as the two founding nations, the officially sanctioned mythology of Canada, "Canada has fifty-five founding [First] nations. . . ." (Dickason (1992, 11)). For the purposes of introducing readers to the philosophies, the teachings, and practicalities of Aboriginal education, and in keeping with the significance of the local, in this chapter we can engage with only a few of the many education-related policies, programs, and places where Aboriginal thought prevails and learning and healing meet. The reader would be wise never to lose sight of the fact that, although there are some similarities in the guiding philosophies across the broad expanse now called Canada, each Aboriginal nation has specific and local teachings based in relationships with the land, waters, air, and all the beings that live in that space. The bibliography that follows provides many opportunities for a fuller, but still introductory, exploration of the issues and context with which those committed to Aboriginal thought as the basis for education and schooling are working.

Circlework: The Medicine Wheel Teachings

The teaching discussed below presents a way of looking at things in relation to one another. For us, it is a useful way to explore relationships significant to our consideration of Aboriginal education in Canada. The circle is a central teaching of many Indigenous cultures. It refuses the linearity and compartmentalization so common to Eurocentric thought. For example, Euro-Canadian notions of history are based in linear timelines; Western

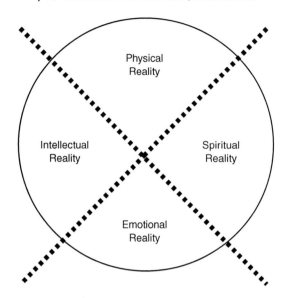

Figure 9.1 Medicine Wheel Showing the Four Aspects of the Self-in-relation

biology teaches us to group living things into kingdoms, phyla, and species. On the other hand, Indigenous thought emphasizes interwoven connections between and among things; life is always in motion. Things exist in ever-changing relation to one another.

Medicine Wheel Teachings, sometimes depicted as a series of interrelated circles that in and of themselves form a larger circle, constitute the underlying pattern (some might say framework) of our chapter. Each Medicine Wheel Teaching in the circle is a comprehensive teaching that connects and relates to the circle on either side. Each Wheel is divided into four quadrants and it is understood that the Creator resides at the centre of those quadrants. The first Medicine Wheel teaching speaks of the four aspects of self (see figure 9.1). Each human being has a spiritual, physical, emotional, and intellectual reality. For the individual to be in balance, each aspect must be in balance; any impact on one reality has an impact on the whole. Few human beings attain this balance in their lives, the pursuit of which is the lifelong pursuit of wellness. Education guides this journey. Too often, schooling focuses on only one or two of the realities human beings live.

The Spiritual Imperative

Chickasaw educator Eber Hampton (1995, 19) writes emphatically that "the first standard of Indian education is spirituality." Hotinonshó:ni educator

Taiaiake Alfred's call for the restoration of Kaienerekowa (the Great Law of Peace) among the Hotinonshó:ni Nations reflects a similar understanding that begins by acknowledging the spiritual and moves outward to others and to the earth. Alfred writes that "the spiritual connections and fundamental respect for each other and for the earth that were our ancestors' way and the foundation of our traditional systems must be restored" (Alfred, 1999, xiv). Alfred's understanding echoes Hampton who relates his Nation's "central prayer as, 'Help me for my people's sake' " or, "Pity me . . . for all my relatives" (Hampton, 1995, 19), while some Anishinaabe conclude their formal speech acts with, "kina nbanwemaa," which roughly translates as "All my relations."

Aboriginal spirituality begins with placing the self with one's relations for the benefit of all of creation. Recalling a fast, Hampton tells of his dawning realization of an expanding consciousness that placed him as part of a greater totality.

> On the second day of the fast, as I prayed I began to ask myself, "Who are my people?" Over the following days my identity expanded from my own skin outwards to family, friends, relatives, Indian people, other humans, animals, growing things, to finally reach the earth itself and everything that is. (Hampton, 1995, 20)

Hampton's expanding awareness of his relationship with all creation is a repetitive theme among Aboriginal peoples and speaks to the connection between land and the spirituality of the people. Willie Ermine contends, "The fundamental understanding was that all existence was connected and that the whole enmeshed the being in its inclusiveness" (Ermine, 1995, 103).

In precontact times, Aboriginal education did not separate spirituality from the learning experience as we do today in schools. We were and are spiritual beings experiencing a physical existence the Creator gives us. That reality still surrounds us, envelopes us, nurtures us, and we in turn can acknowledge that reality through prayer that places us always in relation to our families, our communities, our Nations, and eventually to all creation. Aboriginal Learners and educators attempting to embrace their spirituality as an active part of the educational experience face a difficult task fraught with fear and doubt based in contradiction and incoherence. Mi'kmaq educator Marie Battiste believes that to allow those emotional responses to dictate our actions will "lead us to structures and systems that resemble the old assimilationist models" (Battiste and Barman, 1995, xiv).

It is partly through ceremonies that we can connect to the spiritual reality of an educational journey, which, Ermine writes, "are corporeal sacred acts that give rise to holy manifestations in the metaphysical world. Conversely, it is the metaphysical that constructs meaning in the corporeal" (Ermine, 1995, 106). It follows then that a system of education that forcibly disconnects the teachers and Learners from their traditional metaphysical experience will exist in a corporeal world that has little meaning. Elizabeth Minnick adds that "it is in and through education that a culture and polity, not only tries to perpetuate but enacts the kinds of thinking it welcomes, discards and/or discredits the kind it fears" (Minnick, 2000, 6).

Battiste and Henderson observe the resulting societal wasteland that is a direct result of this metaphysical disconnect created by education as we now practice it. Eurocentric curricula tend to isolate the known self; instead of creating communities, they reinforce specialized interests among students. These curricula teach that knowers are manipulators with few reciprocal responsibilities to the things that they manipulate. These students may know a body of transmitted knowledge or a set of skills, but too few know how to learn or how to live in freedom. Often these students have no inner sense of responsibility to truth and justice (Battiste and Henderson, 2000, 88). Battiste writes:

> ... disconnected from their own knowledge, voices and historical experiences, cultural minorities in Canada have been led to believe that their poverty and powerlessness is the result of their cultural and racial status and origins. In effect, their difference is the cause of their impoverishment state. (Battiste, 1998, 7)

The subconscious but clear message to the dominated who are imparted Eurocentric education is one of cultural inferiority and subservience, which encourages enfeeblement, self-hatred, and anger. Bringing this subconscious legacy of the colonial experience to our conscious minds and discovering that ancient intellectual heritages are more relevant today than ever before should be the goal of all educators with a commitment to Indigenous thought and the well-being of their students. Decolonization of our communities begins with the intellectual pursuits of decolonization and cultural affirmation of our teachers.

All My Relations

A second Medicine Wheel teaching brings Aboriginal education in Canada historically and currently into focus. As is already evident, relationships are central to Indigenous thought. The circle in figure 9.2 with its six directions also draws on the work of theorist Eber Hampton, this time to place four aspects of First-Nations education in relation to each other, as well as (as they always already are) in relation to land and spirit.

As he suggests, it is not a model but an effort to present to the reader a pattern that has directed our exploration of Aboriginal education in Canada. We begin in the east, *Remembering*. Avoiding the studied amnesia

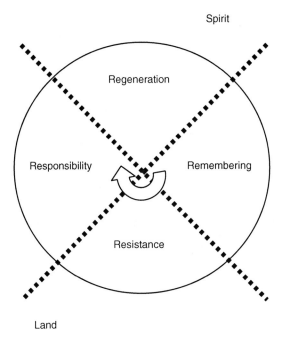

Figure 9.2 Toward Aboriginal Thought in Education, Six Directions (after Eber Hampton)

of many non-Aboriginal Canadians, this direction brings in the generations who came before and calls for a literal remembering in the sense of putting epistemological pieces back together, pieces that have been under assault for centuries, pieces that the knowledge keepers have carried with them. As we move to the south, we find *Resistance*, which has allowed the continued existence and evolution of Aboriginal thought in the face of a persisting colonial presence. Education has at times been a site of resisting curricula, both hidden and explicit, and even opposing schooling[11] in order to maintain the glowing embers of Aboriginal languages and cultures. For some, this resistance has been life-saving; for others, it brings sorrow, pain, and death. *Responsibility* for the generations who came before and the generations who come after, responsibility to keep the teachings current for the good of the all is a focus of Indigenous education. This sense of responsibility connects to the northern door where *Regeneration* of Indigenous thought—language and culture—is always possible and is now in process. As Makere

Stewart-Harawira, a Maori scholar currently living in Cree territory in Canada, tells us, "far from irrelevant in the modern world, traditional indigenous social, political and cosmological ontologies are profoundly important to the development of transformative alternative frameworks for global order and new ways of being" (Stewart-Harawira, 2005). Articulating these transformative, alternative ways of being, these Indigenous ways of thinking and being in the world, in relation to their relevance for broadening mainstream conceptualizations of knowledge, is central to current work in Aboriginal education in Canada. Rather than seeing Aboriginal knowledges as old and passé, we concur with Métis scholar Carl Urion who articulates, "Traditional knowledge is living knowledge."[12] Traditional knowledge has the potential to transform Eurocentric ways of thinking about education and schooling.

Remembering: Recreating the Traditional in Contemporary Context

Let's begin our circle, our discussion, in the East. Traditionally, education for Aboriginal children was rarely separated from everyday life.[13] An integral part of the community, as children grew physically and matured in other ways, they took on individually appropriate responsibilities. To this day, in many contexts, learning tasks by watching and doing continues, with learners generally relying less on questions and answers and more on observation and trial. Speech and the words involved are treasures to be used with care and respect. Storytelling and orality convey the teachings, often focusing on the moral and ethical dimensions of living, always with good humor. Lifelong teaching and learning are assumed: from first efforts of the toddler to protocol in the feast house or longhouse or ceremony, people take up the teachings and continue to pass them along, always *recreating the traditional in contemporary context*, endless human work.

Accompanying the nineteenth-century imperialism that drove Europeans to seek new resources, new labor, and new markets worldwide, schooling was a central tool of colonization. In 1876, the Canadian government commissioned Nicholas Davin to report on industrial schools established for Native Americans in the United States with an eye to their utility for Canada's concerns, seeking a model from our neighbor to the south as Canada is too often wont to do. The outcome of this study was to address the experienced difficulties of "Christianizing and civilizing" the First peoples, and schools were to become the primary tools for this work. Davin's report concluded, "If anything is to be done with the Indian, we must catch him (sic) very young."[14]

Residential schools operated for most of the twentieth century in Canada. At their worst, children were forcibly taken from the families and communities for periods of up to ten or more years. Their languages and cultural practices were forbidden, often with extreme punishment for failure to comply. In an 1891 discussion of the government's determination to "un-Indianize the Indian," a missionary and former residential school principal had this to say,

> How would we white people like it if because we were weak, and another people more powerful than ourselves had possession of our country, we were obliged to give up our little children to go to the schools of this more powerful people—KNOWING that they were taken from us for the *very purpose* of weaning them from the old loves and old associations—if we found that they were most unwillingly allowed to come back to us for the short summer holidays, and when they came were dressed in the peculiar costumes of our conquerors,[15] and were talking their language instead of the dear old tongue, and then—if, when the time stipulated for their education was drawing to a close, and we were looking forward to welcoming them back to the old home, we were to be coolly told that provision had been made for them to go and live elsewhere, and that we were not very likely to see them again? (Emphasis in the original.)[16]

The scars from these schools still affect families throughout Canada. In 2008, the federal government is reluctantly trying to find ways to acknowledge, redress, and apologize appropriately for the horrors wrought. The memories haunt schooling throughout the nation—sometimes recognized, more often percolating below the surface, a nagging reminder of the worst-case educational scenario.

In terms of current education in Canada, *Remembering* must be the starting place for any serious engagement with Indigenous thought. Not only in schools is such work possible: a community can be the site of continuing education. The National Museum of Civilization has a travelling exhibit based in archival photographs of residential schools. Accompanying video and study-guides direct those interested in deeper understandings of these parts of Canadian history. In Red Lake, Ontario, it recently served as incentive for creating a locally based retrospective of the residential school that has operated in the area. Anishinaabe scholar Kaaren Dannenmann and Haig-Brown organized a successful community workshop that used the exhibits as the basis for information and discussion for any interested townspeople. In addition to bringing history alive, the possibility for face to face interaction and storytelling provided a powerful experience for those who participated. Oral tradition is of ongoing significance to Indigenous thought, allowing the passing along of embedded teachings in speech acts accessible in a special way in face to face interactions. Translation to text is a fundamental shift in medium, one that, while powerful and often ironically useful for regeneration of language and culture, is incommensurable with oral tradition. The Supreme Court in 1997 gave official Canadian recognition to oral tradition and oral history—a recognition that has persisted for many Aboriginal people since time immemorial—as it was accepted as evidence in a major land title decision. In schools and communities, when the time is right, elders and other knowledge keepers pass along the stories to all those prepared to listen and act on them.

Resistance: Refusing the Status Quo

One of the most notable things about the imposition of Eurocentric forms of schooling on Aboriginal peoples in Canada is the ability of the most agile to select what works for them and to resist the negative aspects of what is offered. Residential schools with their blatant goals of obliterating Aboriginal languages, cultures, and ways of being began a slow demise in the 1960s. Aboriginal peoples, who had been voicing their opposition for decades, were finally heard. At this point, integration into publicly funded provincial schools[17] took precedence. In addition, attendance at federally funded day schools, which had already been operating in some locations, became an increasing possibility. In 1972, in response to proposed legislation related to the federal Indian Act, the National Indian Brotherhood, a group of very concerned First Nations scholars and activists, prepared a landmark document, Indian Control of Indian Education. Indicating that the transition to provincial schools had been less than smooth for many Aboriginal students, the document focused on the lack of attention given to First-Nations-focused education in the public schools. The report notes, ". . . . it has been the Indian student who was asked to integrate: to give up his (sic) identity, to adopt new values and a new way of life" (National Indian Brotherhood (1972, 25–26)). Calling for local control and parental responsibility as guiding principles for reform, the group spoke out for curricular change, increasing numbers of First Nations teachers and professional development for non-Aboriginal teachers, strongly resisting the insensitive status quo. This focus on teacher education is a recurring theme in Aboriginal education in Canada and provides the primary examples that follow. Other considerations, such as language programs, alternative schools, and curriculum and policy initiatives, could as easily exemplify the tensions to be negotiated and the progress being made today.

This need for educators with knowledge of Aboriginal histories, cultures, and traditions who are capable of responding to the specific needs and interests of Aboriginal students and communities continues to be identified as a priority for schools across the country. More recently, the beneficiaries of such educated teachers are seen to be not only Aboriginal children but also non-Aboriginal students who are expected to develop some understanding of Canada's current and historical relations with

Indigenous peoples. Informing all citizens of our history together has come to be seen as a step in the process of healing ourselves from the colonial relations that continue to shape our realities. As recently as March 2008, *Horizons*, the Government of Canada's policy research journal, was dedicated to Aboriginal Youth and Canada's Future more broadly defined. Their success in schools, often directly affected by the presence of Indigenous teachers with a commitment to Indigenous knowledge, is seen to be integral to the future for all Canadian citizens.

For teachers who take up the challenge of leading their students to remember the resistances that have so often exemplified relations between Aboriginal peoples and other Canadians, the available resources are vast. Films such as those of Alanis Obomsawin, an award winning filmmaker of the Abenaki Nation, portray these conflicts in such films as *Is the Crown at War with Us?*, a documentary examination of a lobster fishing conflict between Mi'kmaq fishers and others, and *Kahnehsatake: 270 Years of Resistance*, which looks at the Mohawks' efforts to protect land in their territory in the 1990s. An award winning feature film *Where the Spirit Lives* portrays one child's experience of residential schools in an easily accessible and moving story.

In another twist on resistance with relevance for those concerned with doing justice to Aboriginal education and Indigenous thought, Susan Dion (2008) (Leni Lenape) analyses the resistance of non-Aboriginal teachers to engaging in the contentious issues that relations between Aboriginal people and other Canadians introduce to classrooms. Stymied by their own ignorance of the context of the biographies of the Aboriginal people whose stories form the basis of their classroom instruction, and conscious of their responsibility to educate good citizens who will see Canada as a country worthy of their loyalty, the teachers frequently deflect difficult questions, failing to engage deeply with the histories and current context of Aboriginal and non-Aboriginal relations.

Responsibility: All Our Relations

Turning to the Western door, we look to our responsibility to educate ourselves in order to work respectfully with the past to serve current and future generations. Alternative ways of knowing in educating peoples, in creating success for those who want to learn, follow two general paths. First is the creation of true alternatives to existing institutionalized education. Second is a recognition of the plea—now more than forty-five years old—for teacher education that requires all teachers to be knowledgeable of the interwoven histories of Aboriginal and non-Aboriginal peoples. We will consider the latter point first and then move to consider the other.

Recent policies, both federal and provincial, are calling for better schooling for Aboriginal people. The recommendations of the 1996 Royal Commission on Aboriginal Peoples (1996a/b/c/d/e/f), echoing the earlier call from the National Indian Brotherhood, included the following recommendations:

> [That] Canadian governments, Aboriginal education authorities, postsecondary institutions and teacher education programs adopt multiple strategies to increase substantially the number of Aboriginal secondary school teachers, including
>
> (a) promoting secondary school teaching careers for Aboriginal people;
> (b) increasing access to professional training in secondary education. . . .
>
> [That] provinces and territories require that teacher education programs . . .
>
> (b) develop options for pre-service training and professional development of teachers, focused in teaching Aboriginal students and addressing Aboriginal education issues. (Section 3.5.18)[18]

In September 2004, the Council of Ministers of Education of Canada (CMEC) identified Aboriginal education as a priority issue. Ministers acknowledged the need to find new and varied ways of working together to improve the outcomes of First Nation, Métis, and Inuit students across both the elementary, secondary, and postsecondary education systems (CMEC, 2005). The following year, the CMEC developed their Aboriginal Education Action Plan. Long-term goals of that plan include positive Aboriginal learning experiences, improved student well-being, increased success for Aboriginal students, and improved labor market attachment[19] for Aboriginal peoples. The first phase of an ongoing commitment to improve education for Aboriginal students identifies the need to "establish mechanisms to recruit and train more Aboriginal teachers" and to "increase the numbers of Aboriginal teachers—those entering as well as remaining in the profession" (CMEC, 2005, 2–3). In 2008, a proposed Indigenous Teacher Education Program (ITEP) at York University is one effort to fill this identified need. It joins a multitude of programs that came into existence in universities across Canada, starting in the early 1970s, focusing primarily on preparing elementary teachers.[20] The new ITEP program is designed to increase the number of qualified Indigenous teachers certified to teach in Ontario secondary schools. Second, the program will also serve non-Indigenous teacher candidates interested in preparing to teach in ways respectful of Aboriginal history and traditions and responsive to the needs of Aboriginal students and communities.

In late 2007, the Ontario Ministry of Education released the Ontario First Nation, Métis, and Inuit Education Policy Framework. The framework identifies Aboriginal education as one of its key priorities, stating: "The overriding issues affecting Aboriginal student achievement are a lack of awareness among teachers of the particular learning styles of Aboriginal students . . . a lack of understanding of cultures, histories and perspectives. . . . a need for appropriate teaching strategies . . . curriculum that reflects cultures and perspectives . . . effective counseling and outreach . . . a school environment that encourages student and parent engagement." In order to achieve successful outcomes, "It is essential that First Nation, Métis, and Inuit students are engaged and feel welcome in school, and that they see themselves and their culture in the curriculum and the school community" (Ministry of Education Ontario, Aboriginal Education Office (2007, 5–7)). One strategy to achieve the desired outcomes is to "encourage faculties of education and community colleges to attract, retain and train more First Nation, Métis, and Inuit students to become teachers and education assistants knowledgeable about their own culture and traditions" (Ministry of Education Ontario, Aboriginal Education Office (2007, 12)). Notably absent is any reference to spirituality, which so many Aboriginal educators see as integral to sound education.

Bachelor of Education in Aboriginal Adult Education

This brings us to our consideration of another teacher education program that serves as one of those "true alternatives" to existing programs through its adherence to Aboriginal thought. It focuses on making Aboriginal teachers of Aboriginal adults. The Bachelor of Education in Aboriginal Adult Education program (ABADED) is offered by Brock University's Faculty of Education through the Tecumseh Centre for Aboriginal Research and Education. Because it is answerable to the communities and ultimately the university and is not designed to license teachers for provincial public schools, this program has more flexibility than those that must answer to the Eurocentric regulations of the Ontario College of Teachers.[21] It is also open to Indigenous thought as a firm foundation and constant source of inspiration. Reflecting the responsibility to remember and to resist colonial forms of education, a learning and healing pedagogy is its unique focus. Teacher candidates are drawn from diverse sectors in the Aboriginal community, including those working with health, youth, housing, friendship centers, and with the police, as college educators, and other Aboriginal service providers. The common links are that each candidate works with Aboriginal adult populations and that some aspect of their work is educational.

Extensive consultation with the Aboriginal community identified administrative and pedagogical elements integral to increasing success for Aboriginal adults (see Kompf and Hodson, 2000). The former included providing advanced standing for college diplomas, university degrees, or course credits, offering the program through distance education in Aboriginal communities, using small classes and a cohort approach, employing local instructors familiar with local issues, and offering classes on weekends. Five core courses are offered in Aboriginal communities across the province of Ontario[22] and include several hours of video of Aboriginal Elders, educators, and other national experts engaging with Aboriginal adult education. Teaching through culture, the videos reflect many Indigenous ideals, including talking circles, connection to the land, and traditional arts and crafts. Candidates travel through the duration of coursework, which is conceived of as an interconnected circle of learning.

Pedagogical Considerations: Learning and Healing

Cognizant of their histories and existing circumstances, many Aboriginal peoples in Canada are heavily invested in healing and wellness as a process of extracting individuals, families, communities, and nations from the contemporary outcomes of the colonial era and reengaging with traditional cultures, values, beliefs, and languages. Incorporating this understanding into education results in what Hodson elsewhere has called Learning and Healing Theory (Hodson, 2004). Learning and Healing recognizes that contemporary Aboriginal peoples have inherited socioeconomic, sociocultural, and sociospiritual realities as a direct result of assimilationist Canadian policies. They may have little consciousness of the impact of the past on the present. Breaking away from these contemporary realities can occur only through a process of consciousness raising, or what Paulo Freire (1970) names conscientization, accompanied by engagement with Indigenous thought embedded in traditional cultures, languages, and spiritual beliefs.

Learning and Healing, as pedagogy and spiritual endeavor, resists human attempts to fully apprehend and communicate in text what those words can mean, how learning and healing transpires, and under what conditions. Learning and Healing rejects deficit theorizing (see Bishop and Glynn, 2003). Learning and Healing is an Indigenous wellness model that understands that all are engaged in a healing journey and that the conduct of that journey to a large extent is within individual control. One way to approach a deeper understanding of such an approach to life is through traditional ceremonies. The Sweat Lodge, the Sun Dance, and the Vision Quest followed by many Nations or the Hotinonshó:ni Condolence Ceremony (see Alfred, 1999) all require a conscious test of the physical and material selves to seek spiritual insight and restore the balance within and in the world around us. If in these sacred endeavors one's motivations are pure and worthy, one will receive assistance from the spirit-world, but that assistance is rarely provided in the way that is expected. When immersed in a process of Learning and Healing, one has an opportunity not only to create a relationship with oneself but also with others who are likewise engaged. These heart-to-heart relationships have the potential to expand organically into mutually supportive communities[23] of like-minded people who can flourish and continue their own healing journeys.

This environment also provides opportunities for students and teachers to begin or continue the process of decolonization, of gaining knowledge and an understanding of what they have experienced and how it has shaped us all today. We may connect with and deconstruct our historic and contemporary realities to discover why we are the way we are and how we have incorporated particular colonizing values and beliefs into our lives. At the same time, we gain insights that encourage us to leave behind alien values and beliefs and to discover a new relevance and confidence in traditional values and beliefs, which we may begin to incorporate into our everyday lives. As a result, we can move through our lives in a different manner, making different choices, and demonstrating different behaviors; some refer to this as "walking the good red road." That expanding wholeness affects the world and creates change, although we may be unaware of that occurring.

Learning and Healing recognizes that this process of discovery begins with an unmasking of the impacts of colonialism on peoples, their families, and communities. The role of the Aboriginal adult educator is to facilitate this process by cocreating an environment that addresses the spiritual, emotional, intellectual, and physical. Only then will individuals have the opportunities for reflective analysis that encourages new understandings and new ways to move closer to wellness. Learning and Healing is not new educational theory, nor should it be considered emerging knowledge. Learning and Healing is the contemporary expression of traditional Indigenous forms of education that were conceived as a way of promoting and maintaining balance in the individual, the family, the community, and the Nation. To be sure, such an integration and the associated actualization are not homogenous in nature and vary widely depending on the capacity and circumstance of the individual. The reality of Learning and Healing is closely related to the experience of birth, with all its associated blood, sweat, and tears, where the contractions that precede birth are not consistent or constant but arrive in waves dictated by circumstances beyond control or prediction. Each individual must negotiate this reality in his or her own way. The Learning and Healing journey is an organic experience, chaotic and dynamic.

Regeneration

Remembering the past, resisting assimilationist schooling, taking responsibility for selves and the education of all our relations, those committed to Aboriginal education in Canada today are creating sites for regeneration of Indigenous thought and all that it implies. Increasingly, Indigenous thought in its many iterations not only informs Aboriginal people but also reaches into non-Aboriginal people's lives and shapes their realities. In this section, we introduce several concrete expressions of Aboriginal thought in action and ponder some directions for research that brings Indigenous thought to bear on current education, especially that involving all those who live in Canada.

Starting from the fundamental understanding that people construct knowledge, we claim that they can therefore deconstruct, reconstruct, or regenerate it. For our purposes, the word regeneration names that which is constantly being recreated even as it can never fully escape its history—endlessly spiraling down the generations into the future. When what have been distinct knowledge systems, as evident in cultures and their inextricably linked languages, encounter one another, Homi Bhabha (1994) tells us that the ensuing result is a new and contested space drawing on each other but never the same as what came before for either system. In Canada, where Aboriginal knowledges at least figuratively and historically underpin all of what we are, they are a reasonable place to reimagine what our relations to one another have been, might be, and could become.

What does regeneration of Indigenous thought look like in this context? It looks like the Native Indian Teacher Education Program, now over thirty years old, and the Ts'kel program for educational leaders at the University of British Columbia; it is embodied in the First Nations House of Learning there, in a multimillion dollar, magnificent longhouse. It is present in the PhD in Indigenous Peoples Education at the University of Alberta, which includes sharing circles, Elders' support, and ceremony.[24] It is present in the Aboriginal Literacy Education Certificate through the Office of First Nations and Inuit Education at McGill University in Montreal, a certificate that teaches literacy to fluent speakers of an Aboriginal language. It looks like language immersion schools from Chief Atahm School in Sexqeltqin in Chase, British Columbia, a Secwepemctsin Immersion program within the Secwepemc Nation, to Kawenni:io/Gaweni:yo High School, a private secondary immersion school in Six Nations, Ontario, offering a curriculum that is 40 percent in either Mohawk (Kanien'keha:ka) or Cayuga (Gayogoho:no) and the associated Kawenni:io Elementary School.[25] It is exemplified in plans to rebuild the Mi'kmaq immersion program in Elsipogtog, Nova Scotia, where it has suffered from a lack of funding, a struggle many First Nations initiatives face.[26]

In terms of policies, directives related to the regeneration of Aboriginal education are expressed in documents such as New Brunswick's 2008 tripartite Memorandum of Understanding on First Nations Education, which announces $70 million to improve First Nations education;[27] Saskatchewan's *Aboriginal Education Provincial Advisory Committee's Action Plan* for 2000–2005; the Aboriginal Education Office of the Ministry of Education in its

Ontario First Nation, Métis, and Inuit Education Policy Framework of 2007; and the Manitoba First Nations Education Resource Centre with its stated goal "to help all First Nations improve education for all learners to achieve."[28] A focus on Aboriginal education reminds us of the rights to education guaranteed to those First Nations who have signed treaties with Canada.[29] Regeneration of Indigenous thought is the basis for the annual imagineNative Film and Media Arts Festival in Toronto and the IMAGeNATION Aboriginal Film and Video Festival in Vancouver: education broadly defined. Each year, the National Aboriginal Achievement Awards honor Aboriginal people recognized for outstanding accomplishments in various disciplines including health, law, political science, culture, and arts. Indigenous thought is represented in many of the texts included in our bibliography.

In the face of all of this activity, we must remember to remember even as we look to regeneration. We recognize that living knowledges are never fixed in time but are constantly evolving. Anishinaabe author Louise Erdrich (2004) writes, "[E]ven a people who . . . were saved for thousands of generations by a practical philosophy, even such people as we, the Anishinaabeg can sometimes die, or change, or change and become" (210). Sites of education provide one place where we all might begin consciously to "change and become" through working to decolonize our lives by recognizing what we consider our histories to be and what meanings we make of those assumptions. Regenerating languages and cultures is not without its difficulties. As Marlene Brant Castellano notes, Aboriginal education is a ". . . landscape in which hope and possibility live side by side with constraint and frustration" (Castellano et al., 2000, 251). Aboriginal people encounter tensions as they work with Indigenous knowledge in contemporary contexts, especially in relation to what are seen as the traditions that underpin it and its current iterations after 500 years of colonization on this continent, and especially in work with non-Aboriginal peoples. Arguments for the irrelevance of Indigenous knowledge in this time of market economy and global expansion of those markets, dismissive comments about romantic desires to return to the "old days," and accusations of cultural appropriation mediate against those who would take Indigenous knowledge seriously.

Our educational research questions take up the nuances and intricacies of the meeting of cultures and the paradoxes of their persisting and ever-changing realities. We query the relation between the regeneration of Indigenous thought and current and future education. What is its significance for cultures, for Aboriginal cultures in the Canadian context, and for non-Aboriginal cultures? If one takes seriously the notion that culture becomes explicit and is always affected by its contact with another culture (Wagner, 1981), what happens when students and teachers engage with Aboriginal knowledges in places like schools and whitestream[30] communities where

European-based cultures and languages have predominated for generations (Willinsky, 1998)? What happens when students and informed teachers in these schools and other sites of education focus on accessing cultural traditions (regenerating old ways) even as they are creating fresh hybrid culture (generating new ways) for thinking and being in the world?

As we research these questions, we draw on the work of scholars of Indigenous knowledges engaging in decolonizing work (for example, Battiste, 2000; Stewart-Harawira, 2005; Tuhiwai Smith, 1999) and theorists who focus specifically on education as a colonial practice (for example, Foucault, 1979; Smith, 2006; Willinsky, 1998). Why decolonizing? Our work takes seriously the understanding that Canada continues to be a country whose roots lie in colonization (Ng, 1993; Loomba, 1998) and that there is much unfinished business in this regard. Decolonizing work, Tuhiwai Smith (1999) tells us, is work with a purpose: to improve people's lives through demystifying knowledge production by situating Indigenous knowledges in their rightful place in our consciousness where they may serve to contribute views alternative to those arising directly out of the Enlightenment in Europe. We work not only to regenerate and name new forms of old knowledge but also to create the possibility of generating new ways of thinking through relations with one another.

In our research, we consider the effectiveness of educators' efforts to decolonize their educational practices. We go to real sites of teaching and learning that attempt to take Indigenous thought seriously as we seek to evaluate the development and implementation of pedagogical approaches for success for Aboriginal students; to assess the development and implementation of decolonizing workshops and other educational events, including their effects as regenerative processes; and to contribute to theorizing the relations between and among Indigenous thought, other knowledges, and the process of decolonizing. On the other hand, we resist decolonizing theory when it becomes too strong a focus and Indigenous thought is relegated to conflict and struggle rather than being taken up on its own terms starting from a complex engagement with those four interrelated realities: the physical, emotional, intellectual and spiritual. We take seriously the dismal and deadly statistics of Aboriginal students' school leaving, drug and sexual abuse, representation in prisons, and suicide. We resist the knowledge that tells those young people that they are wrong, and are unwelcome in Canadian society. We seek ways to restore their souls, and ways to educate their teachers.

In the space of interrogation and analysis that academe allows, contesting ways of being in the world and ways of thinking about the meanings people ascribe to things collide, interact, and transform one another. These interactions occur in schools, in other sites of education such as community initiatives, as well as in everyday informal settings. We seek engagement with, and renewal based in, Indigenous thought in all its complex and evolving iterations as a way to consider the possibilities, indeed the inescapability, of regeneration of knowledges. Simultaneously, we ponder the limitations of meaning-making when people are immersed in their own contexts. To address the limitations of immediate experience and to deepen understandings of the ways that people are making sense of their experiences, some of us read and ponder the work of scholars of Indigenous thought in relation to other decolonizing and postcolonial work. Ultimately we want to contribute to generating alternative ways of thinking through what it would mean and look like, both discursively, practically, and policy-wise, for Canadian education to take Aboriginal knowledge seriously. Here lies further possibility for recognition of the strength and contributions that Aboriginal knowledges have made and continue to make to our common communities and the nation. Indigenous knowledges, in their insistence upon a divergence from what David Smith (2006) has called the Enlightenment's goal of a singular (un)truth, have much to contribute to generating new alternative discourses. The success of Aboriginal education in Canada rests in this new goal. All my relations.

Notes

1. Eber Hampton (1995, 42).
2. Marie Battiste (2000, xvi).
3. The government of Canada passed the Canadian Charter of Rights and Freedoms in 1982. The Constitution recognizes the rights of Aboriginal peoples of Canada, which include Indian, Inuit and Métis groups. For the purposes of this chapter the authors will identify various educational initiatives through these categories and wherever possible by a specific tribal or national affiliation. The term First Nation identifies Indian communities while Indigenous is used as a global term.
4. We acknowledge the danger of collapsing many diverse forms with this category as well as the irony of doing so in light of our arguments. However, for the purposes of this short discussion, the word will stand in as that which is oppositional to Indigenous thought in a North American context.
5. Hampton writes of ". . . mother earth, the sustainer and source of rebirth." (1995, 19).
6. Although the term Aboriginal is used throughout this chapter, readers are cautioned to recognize that this term does not signify or imply a common reality for all Aboriginal peoples who live in this northern nation. The names people have for themselves, for example, Secwepemc, Mi'kmaq, Anishinaabe, Hotinonshó:ni, Inuit, and Métis, better capture the highly diverse cultures, languages, values, beliefs, and histories involved. Within each nation, there are, of course, tremendous individual variations in contemporary realities and aspirations.

7. For those who worry about a religious undertone with the use of "spirit," Eber Hampton may provide some solace or even food for thought. He associates spirit with "the great mystery" (1995, 19). One might also connect it to a humility that simply acknowledges the limits of our knowing.
8. In another related context, seahorses are back in the Thames after over thirty years dedicated to cleaning that river. *Seahorse found in Thames Estuary,* June 14, 2004. http://news.bbc.co.uk/2/hi/uk_news/england/essex/3806593.stm. Accessed 04/16/08.
9. As immigration continues apace, there is an urgent need in Canada to educate current citizens and recent immigrants of the history of colonization and the outstanding, legal and moral land claims still to be resolved and Aboriginal rights still to be recognized.
10. In Nunavut, the newest territory, for example, its capital Iqaluit was for a time named Frobisher Bay.
11. The distinction between education and schooling is an important one. Education refers to all the teachings and learning that transpire between and among people in any context; schooling refers to the institutionalized forms of education, such as the teaching and learning that happens in schools, universities, and colleges.
12. Carl Urion, cited in Stewart-Harawira (2005, 35).
13. At specific life-changing times such as puberty and childbirth, separation of the person involved from aspects of the everyday might occur for the purposes of ceremony and recognition as well as teaching.
14. Cited in C. Haig-Brown (1988, 30).
15. It is important to note the words of Law Professor John Borrows (Anishinaabe) of the University of Victoria in his discussion of the treaties signed in many areas of Canada, "A treaty is not a surrender." (From Haig-Brown's notes of a lecture given at York University in the late 1990s.)
16. Cited in Nock (1988, 165).
17. In Canada, fiscal responsibilities are divided between the federal and provincial governments. The federal government is responsible for Indians as defined by the Indian Act, a separate body of laws pertaining to specified individuals of Aboriginal ancestry. The provincial government is responsible for public education. This results in tuition fees being paid for "Indians" to attend public schools.
18. Cited in Castellano et al. (2000, 259).
19. While this goal might be considered with some skepticism as a continuation of the goal of residential schools for creating workers for the country, the dire statistics related to Aboriginal poverty indicate the importance of acquiring scholarly skills, abilities, and access to jobs.
20. In Canada, the schools tend to be divided into primary grades that include kindergarten and the first three years of school (ages 6–9), elementary or junior grades that range from four to seven (ages 9–12), and secondary schools that include grades eight to graduation following grade 12 when students are 18 or 19 years old.
21. "The Ontario College of Teachers is the largest professional self-regulatory body in Canada with 200,000 members. It was founded in 1996 to license and regulate teaching in the public interest. The College ensures the province's

teachers are qualified and competent, and that students are safe in their care." http://www.enseignerenontario.ca/en/oct.htm. Accessed April 23, 2008.
22. Ontario (from the Iroquois word Kanadario, meaning land of shining waters) is the second largest province in Canada with over a million square kilometers. There are 143 Aboriginal communities and a population of approximately 250,000 people living in these communities and urban centers. Diversity rules!
23. We fully acknowledge the fleeting and difficult moments that constitute community. However, these difficulties never keep us from seeking those moments of healing together. See for example, Haig-Brown and Dannenmann (2002).
24. http://www.uofaweb.ualberta.ca/education/aboriginalprograms.cfm. Accessed April 24, 2008.
25. http://www.icmi.ca/immersion/immersion.html. Accessed April 23, 2008.
26. http://telegraphjournal.canadaeast.com/front/article/275816. Accessed April 23, 2008.
27. http://news.gc.ca/web/view/en/index.jsp?articleid=393739&categoryid=7&category=Regional+News. Accessed April 24, 2008.
28. http://www.mfnerc.org/index.php?option=com_frontpage&Itemid=39. Accessed April 23, 2008.
29. Federation of Saskatchewan Indian Nations. Treaty Right to Education. A Presentation Report from the Federation of Saskatchewan Indian Nations, Education and Training Secretariat. No date.
30. After Sandy Grande (2004).

Chapter 10

Montessori and Embodied Education

Kevin Rathunde

Introduction

Montessori education has been around for over 100 years; however, it still has much to contribute to a new vision of twenty-first-century education that supports well-rounded human development in democratic societies. A main theme in this chapter is that emerging interdisciplinary thought on the embodied mind has provided a new framework for understanding Maria Montessori's contributions to education. Montessori philosophy, I believe, offers an alternative approach that might be called *embodied education*: education in tune with the intimate connection of the body and the mind. Such coordination of body and mind is important for education because it facilitates student experiences of deep engagement and interest that have been referred to as *flow* (Csikszentmihalyi, 1990). Flow-like experiences, in turn, have been associated with intrinsically motivated learning and talent development (see Csikszentmihalyi, Rathunde and Whalen, 1997).

The chapter is organized in three parts. The introductory section suggests why alternative visions of education are needed. It argues that many traditional schools have drifted toward *disembodied education* where students are drilled cognitively and suffer the experiential consequences of drudgery, lack of motivation, and mental fatigue. The second part of the chapter articulates the distinctive features of Montessori education and suggests how they contribute to holistic human development. To help illustrate the key ideas of Montessori education, close parallels are drawn to optimal experience (flow) theory (Rathunde and Csikszentmihalyi, 2006). Finally, part three summarizes recently completed research supporting the claim that Montessori education is associated with student engagement and flow.

The Drift Toward Disembodied Education

A great threat to the quality of student experience and motivation comes from disembodied education, or education that is overly focused on the cognitive or abstract part of the learning process. In other words, disembodied education takes the body, and the activity, movement, and emotions associated with it, out of the mind. It operates on the misguided assumption that the mind exists out of context, as if suspended in an objective state. Modern neuroscience (see Damasio, 1994) and other disciplines (see Lakoff and Johnson, 1999) have proven this assumption false; nevertheless, the view of the mind as a neutral container—into which important facts and ideas can be poured—continues to have a powerful influence on many educational practices.

In a superficial sense, everyone recognizes that education is never completely disembodied. Even if sitting motionless and reading, the eyes and other parts of the body (that is, the brain) are necessarily involved. However, there are degrees of disembodiment; and some of the most extreme examples are found in traditional schools where students are expected to sit quietly, listen to lectures, do homework, and focus only on the symbols and concepts they are manipulating. Educational approaches that adopt this style too exclusively often damage student motivation. Students in these one-dimensional environments report very little engagement with what they are learning in comparison to other school settings (for example, extracurricular activities) that provide more hands-on and active opportunities (Csikszentmihalyi, Rathunde and Whalen, 1997).

How did we arrive at this alienating point of disembodiment in so many educational environments? Part of the answer is simply that education reflects broader themes that have developed in Western culture. Many scholars have been exploring how the highly rational and scientific culture of the West has narrowed conceptions of human nature and learning. One theme that cuts across many of these accounts is that the body's influence on the mind is so pervasive and taken for granted that the mind is often and easily overlooked (Leder, 1990). We look at the world and focus on what we see, but we cannot see the eyes that do the seeing. Even the internal rhythms of our bodies (for example, our heart rate and respiration) affect our perception, but they are so automatic that they are relatively invisible and seldom enter consciousness (see Gallagher, 2005).

Although it is one of humankind's greatest achievements and source of much of our intellectual power, it must also be acknowledged that language plays a key role in disembodiment. The advent of the alphabet and the ability to represent objects with the written word has an unrecognized potential to seal us off from our sensory environment and create a relatively self-referential system of meanings (Abram, 1996). Language gives us a seeming independence from our bodily existence; it allows us to interact with our own signs and symbols and to be completely immersed in human channels of communication such as books, television, and computers. As these technologies have grown, so has the potential risk of disembodiment.

Abram's (1996) linguistic analysis suggests that a consequential step toward disembodiment was taken when the phonetic writing system was adopted by Greece. Socrates and Plato were teaching when this new technology of reading and writing was spreading rapidly, and their foundational insights of Western philosophy were the "hinge on which the sensuous, mimetic, profoundly embodied style of consciousness proper to [oral traditions] gave way to the more detached, abstract mode of thinking" (Abram, 1996, 109). Furthermore, Plato's emphasis on universals that were disconnected from the natural world helped to structure thought in the West and eventually resulted in the foundations of modern science and Descartes' separation of mind and body. Along with this separation came the view that the body was less important, and sensory information and emotion were potent sources of illusion.

It would be a mistake to think that how philosophers and scientists view the world is irrelevant for the practical realities of child development and education. The fact is that leading ideas in a culture influence parents' and teachers' beliefs about human nature; and these beliefs can profoundly affect the way children are perceived and treated. Just as sexist attitudes can affect the day-to-day treatments of males and females, the container-view of the mind that is so ingrained in modern society affects the daily practice of education. The metaphor lends itself to a production-line view of schools and has contributed to the prevalence of disembodied practices that overemphasize the most efficient transfer of facts and bits of information.

Modern science has only recently started to self-correct and recognize the embodied mind. The neurologist Antonio Damasio comments, "The mind is embodied, in the full sense of the term, not just embrained" (1994, 118). His research suggests that the body and emotion are indispensable for rationality and the highest reaches of decision-making and creativity. The insights of many thinkers in child development have also helped to fill out the picture of how we think and learn. Piaget's insights about the sensorimotor period have highlighted the importance of movement and play for cognitive development. Werner's (1956) analysis of the development of language suggests how the use of symbols originally develops out of bodily, motoric, and gestural processes (see also Crain, 2000). In other words, learning language is not simply accomplished by naming things, adopting conventions, and dropping them into the "container." Instead, making reference to things first involves the use of the hand in pointing; an infant's first symbols are often motor imitations (for example, fluttering eyelids to depict flickering lights); and the first vocalizations emerge out of bodily emotional patterns (for example, cries).

While early childhood education has started to benefit from these insights, education in middle childhood and adolescence often disregards the embodied foundations of learning. With age and development, the "geometric-technical" categories of our language and scientific worldview (see Werner, 1956) are increasingly given precedence. Some educators are starting to realize that the intrinsically motivated learning that takes place in early childhood classrooms too often turns into the fragmented, competitive, and extrinsically motivated drudgery of middle- and high school classrooms (Eccles et al., 1993). A number of secondary schools in the UK, for example, have started to implement accelerated learning techniques (see Smith, Lovatt, and Wise, 2005) that emphasize multisensory approaches and the importance of emotional engagement for effective learning. Such attempts to reform education, however, are not widespread. Consequently, many children develop and mature in disembodied educational contexts and progressively lose the ability to tune in to an intuitive, immediate perception of an event or situation. With this loss comes a loss of creativity. Creative work in the arts or sciences requires retaining perceptual/intuitive awareness *along with* the abstract capacity to transform an experience and express it on a symbolic level (Martindale, 1999).

The point of these introductory observations on disembodied education is not, of course, to disparage the use of language or abstract thought in schools. It is, rather, to suggest how misunderstanding human nature and the process of learning can have unforeseen and negative consequences. In an era where video technologies are pushing disembodiment to even greater extremes, it is important to keep in mind what can happen to the quality of life and learning when the embodied foundations of our symbols are forgotten and the symbols themselves are elevated to an untouchable status. This is the case in too many educational contexts; and the numbing effect of feeding dry facts to students not only takes an experiential toll but undermines a student's love of learning.

Disembodiment happens at all levels of education. In early childhood, just as language is exploding and children are beginning to use words to represent the objects they encounter in the world, many teachers, with the best of intentions, take a "name and move on" approach. In middle childhood, when students are moving beyond the immediate environment and using their emerging logic to grasp the broader world, many teachers adopt a "classify and move on" approach. Then again in adolescence, when students have a new capacity to understand what it means to be a social being and have a sense of justice and personal dignity, many teachers just "lecture and move on." These are hallmarks of disembodied educational approaches.

Disembodiment is a problem that Maria Montessori understood well. She cautioned teachers if they followed standard practices that separated body and mind they would become "vivisectionists of the human personality" (Montessori, 1989, 11). She continually emphasized the importance of children's immediate experience in every developmental phase, starting in early childhood with sensory experience. Even writing was taught through the use of sandpaper letters to bring the hands and the sense of touch into the process. For the same reason, Montessori continually emphasized children's feelings of interest and wonder because such states animated intelligence and allowed the mind to accomplish intellectual work.

By emphasizing an immediate, felt connection to the body, Montessori anticipated insights of contemporary work related to embodied cognition. For instance, Abram (1996) argued that direct encounters with the world in an immediate present are needed to creatively expand our language categories. Such forms of participant awareness provide both protection and foundation for our rational intellect: "When reflection's rootedness in such bodily, participatory modes of experience is entirely unacknowledged or unconscious, reflective reason becomes dysfunctional, unintentionally destroying the corporeal, sensuous world that sustains it" (Abram, 1996, 303). Mark Johnson (1987) expresses a similar point of view in his book *The Body in the Mind*: body-based, aesthetic encounters with the environment are important for the expansion of thought. Like Montessori, these authors suggest these are overlooked, embodied dimensions to human learning and creativity.

Montessori Education Reconsidered

Montessori philosophy has been familiar in education circles for many years. It is practiced in thousands of schools worldwide and a growing number of private and public schools in the United States. Some aspects of Maria Montessori's work and perspective have been well documented (see Lillard, 2005; Loeffler, 1992; Standing, 1984; Wentworth, 1999), yet her ideas about how connecting body and mind influenced students' attention and experience have largely gone unrecognized. These experiential insights of Montessori will be highlighted next by drawing parallels between Montessori philosophy and optimal experience (flow) theory. The comparisons between the two perspectives will also shed light on how embodied educational approaches can positively affect the quality of student experience.

The Centrality of Deep Concentration

Montessori placed a central emphasis on students' episodes of deep concentration. This emphasis is the strongest link between Montessori philosophy and flow theory. Montessori believed that children's spontaneous and deep concentration was the natural state of childhood and the essence of being human. E. M. Standing (1984, 174), the authoritative biographer and colleague of Dr. Montessori, called concentration the "key that opens up to the child the latent treasures within him." There is little doubt that what Montessori had in mind when speaking about concentration was something akin to flow. She frequently commented on the single-mindedness of children's powers of concentration: "It has been revealed that children not only work seriously but they have great powers of concentration. . . . Action can absorb the whole attention and energy of a person. It valorizes all the psychic energies so that the child completely ignored all that is happening around him" (Montessori, 1946, 83–84).

Montessori's emphasis on concentration so deep that one becomes unaware of surroundings calls to mind three decades of research that has explored flow and identified its primary phenomenological attributes (Csikszentmihalyi, 1990). In addition to being oblivious to distraction because of the merging of action and awareness, the full concentration of flow is also known to be associated with a change in the awareness of time (for example, time passing quickly), feelings of clarity and control, a lack of self-consciousness, and feelings of intrinsic motivation, or doing an activity primarily for its own sake even if the activity has other extrinsic rewards associated with it. The flow experience is triggered by a good fit between a person's skills in an activity and the challenges afforded by the environment. Therefore, flow always refers to a bidirectional *relationship* to the environment wherein a person is fully engaged with some challenging task. Flow can occur while doing a variety of activities, and people from many different cultures report the experience (Csikszentmihalyi, 1990).

Montessori referred to children who possessed habits of deep concentration (that is, those who had repeated flow experiences) as *normalized* (see Kahn, 1997). "Normal" in this context had nothing to do with what was average for humans. In fact, Montessori thought many adults had often lost the ability to become completely immersed in their activities. Although Montessori's terminology is somewhat confusing, her choice of term "normalized" reflected the fact that she began creating her method during an appointment as a directress at the Orthophrenic School in Italy that served children regarded as hopelessly deficient. Her terminology reflected her belief that problems of attention (for example, distraction, inattention) would disappear along with the first episodes of deep concentration, and if school conditions allowed children to exercise their inborn capacities, spontaneous concentration was the normal or natural disposition of the human organism.

Montessori realized the significance of concentration through observation. A key turning point is described in Standing's (1984) biography.

Montessori was watching a three-year-old child occupying herself with graded wooden cylinders that had to be fitted into a wooden block. The child demonstrated amazing concentration that seemed to isolate her from the surrounding environment. Even when Montessori asked the teacher to have the other children sing aloud and make distractions around the child, her concentration did not falter. Then, as if "coming out of a dream," the child stopped what she was doing, looked around, and appeared happy and at peace. Witnessing this episode evolved into the main theme of the Montessori method: an emphasis on uninterrupted and deep concentration in the classroom. Such episodes were seen as the key to education that was directed, not by external forces in the environment but by what Montessori referred to as "delicate inner sensibilities intrinsic to life" (Montessori, 1981, 252). Today these sensibilities would likely be called intrinsic motivation (see Deci and Ryan, 1985).

The above anecdote reveals to what extent Montessori's entire method had its basis in an experiential orientation toward education. This fact, however, is not widely recognized or discussed in the literature on Montessori. The anecdote also provides the most compelling evidence for the claim made here that flow theory and Montessori education have key ideas in common. Many characteristics of flow that have been revealed by research, including feeling refreshed and at peace on emergence from flow (see Csikszentmihalyi, 1990), were clearly manifested by the young girl manipulating the wooden cylinders.

Why is flow and deep concentration important for education and development? Experiences of total concentration are intrinsically rewarding and motivate a person to repeat an activity that produces them. Repeating the same activity at the same level of challenge and skill, however, will not consistently reproduce the experience. A person quickly becomes habituated to whatever challenge originally engaged them. To recapture flow, therefore, a person must continually raise their level of challenge and/or skill. For instance, a young student who likes science cannot read the same text over and over, or their attention and motivation will fade due to habituation. What first engaged the student must lead to fresh insights and discoveries that expand upon the original material. At first, these efforts to expand knowledge might not produce flow, but as comprehension catches up, and skills and challenges become more balanced, flow occurs again. A great deal of research has demonstrated that such "optimal" experiences are associated with the development of talent and creativity (Csikszentmihalyi, 1996; Csikszentmihalyi, Rathunde, and Whalen, 1993; Csikszentmihalyi and Scheider, 2000).

The developmental dynamic of skills and challenges described above has held a prominent place in psychology throughout the twentieth century, but it has seldom been viewed through an experiential lens. In other words, many influential theorists have proposed such a dialectic *integration* and *differentiation* underlying human development and learning (see Baldwin, 1906; Dewey, 1910; Piaget, 1962; Werner, 1958). What distinguishes flow theory, and other experiential approaches such as Montessori's, however, is the focus placed on momentary subjective experience. While many past perspectives have described the integration and differentiation of knowledge from the "outside" (for example, in terms of various learning outcomes), experiential perspectives reorient the discussion to how the person is connecting with the environment *at the moment* as evidenced by the quality of a person's attention, concentration, and overall experience (see Rathunde and Csikszentmihalyi, 2006).

Preparing a School Environment for Deep Concentration
and Flow: Three Key Balances

Maria Montessori held a holistic view of optimal experience like other scholars who have studied the phenomenon. Her descriptions (1966) of normalization, for example, suggest that an ideal student is joyfully immersed in the present moment *and* capable of using the mind for abstraction and creating order.[1] A flow experience is what Dewey called an integral experience—one that combines affective and cognitive elements at the same time (Dewey, 1934); it is an ideal mental condition of being "playful and serious at the same time" (Dewey, 1910, 218). Abraham Maslow's (1968) qualitative studies of peak experience described a combination of emotional and intellectual synchrony in peak moments. Using the Experiential Sampling Method (ESM), a paging technique that uses watches to signal participants (see Csikszentmihalyi, Rathunde and Whalen, 1997), recent empirical studies have confirmed that a unique combination of affective and cognitive elements come together in episodes of flow.

Given the affective-cognitive nature of the engaging experiences one is trying to facilitate, how does one go about setting up an appropriate school context? Montessori's answer was the *prepared environment* (that is, prepared for student concentration). There are three aspects of a prepared environment that will be discussed next: finding the appropriate balance of skills and challenges; freedom and discipline; and body and mind. The former two balances are important to Montessori, but they do not clearly distinguish her approach from other educational approaches. Therefore, they will be given brief attention before exploring in detail the third balance that is aligned more closely with the theme of embodied education.

Balancing skills and challenges. The first important aspect of a prepared school environment is finding the appropriate balance of skills and challenges. Montessori intuitively understood how this balance influenced the quality of a student's concentration and subjective experience. Therefore, she paid close attention to observing students and selecting activities that were suitably challenging for children at different stages of development. This discovery-oriented approach resulted in her many pedagogical innovations centered on developmentally appropriate activities and sensitive periods (Loeffler, 1992; Wentworth, 1999). Montessori education follows three "planes" of education: birth to age 6, 6 to 12, and 12 to 18. Each plane or level is focused on activities and tasks that find a good fit with students' skills and interests.

Finding a good fit is obviously important to finding flow. When skills outpace challenges, the experiential consequence is boredom; when challenges exceed skills, one experiences anxiety (Csikszentmihalyi, 1990). Many educational approaches, especially those informed by the developmental insights of Piaget and Vygotsky, pay close attention to this same balance; however, Montessori emphasized the phenomenological consequences of a good or poor fit. Montessori teachers were trained to become astute observers of a child's subjective experience (Montessori, 1989). If a teacher could not detect when a child was bored or anxious, it would be impossible to maintain an environment that was suitable for deep concentration. This art of observation, when working well, differs from traditional approaches that view teacher observation primarily as a way to ensure that students are staying on task and finishing assignments.

Because a premium is placed on concentration, one of the primary functions of a teacher in a Montessori environment is to know when to interrupt and when not to interrupt a child. By thinking experientially, Montessori anticipated contemporary research that explores how praise and rewards can encourage extrinsic rather than intrinsic motivation (see Deci and Ryan, 1985). Montessori remarked that a teacher must encourage a child but be careful "not to spoil the perfect dose" such that the child is motivated to work "so that it may obtain merit from her" (quoted in Standing, 1984, 310). She understood that a simple comment such as, "how nicely you are doing that," or even observing too closely what a child was doing, could be enough to disturb concentration. She added (quoted in Standing, 1984, 310), "If a child begins to work with the motive of obtaining praise from us . . . he will begin to develop all sorts of tricks. . . . In this way we might waste that precious energy that is in him."

Montessori also warned against helping children complete tasks they could finish themselves. She believed in a "Golden Mean" where a teacher provided *just enough instruction*, and no more than the indispensable minimum. Standing (1984, 311) summarized her stance on this issue, "The general rule is that the teacher should not intervene when she finds the child engaged in some spontaneous activity which is orderly and creative." Similarly, a teacher using a Vygotskian approach protects a child's motivation and encourages feelings of mastery and efficacy by knowing when to help the child and when to back off and let the child work alone. Guided participation is part of a delicate give-and-take process that pays particular attention to challenging a child at the edge of their skills, or in their zone of proximal development (Rogoff, 1990). From an experiential perspective, it is in the zone of proximal development that a student is more likely to find flow and a synchrony between affect and cognition.

Balancing freedom and discipline. A second key aspect in a prepared school environment involves the balance of freedom and discipline. Perhaps the most widely recognized and discussed element of Montessori education is the freedom students are given to choose activities (Montessori, 1989). Montessori understood that to engage children's emotions and spontaneous interest, freedom of choice was a necessary precondition. A teacher, therefore, tries to create an environment that is in tune with a child's interests. The method is indirect in that it provides materials that a child can actively explore, and often introduces them with a demonstration rather than a lecture. The teacher tries to avoid teaching a child directly with abstract concepts. Therefore, a great deal of time is spent preparing and organizing materials that can be presented to the child in an orderly and progressive fashion.

Unlike other activity based methods that recognize the importance of intrinsic motivation and mistakenly provide children with unlimited freedom, *Montessori never lost sight of the opposite and equal need for order, structure, and discipline in the environment.* She commented, "On this question of liberty . . . we must not be frightened if we find ourselves coming up against contradictions at every step. You must not imagine that liberty is something without rule or law" (quoted in Standing, 1984, 286). Freedom and discipline are seen as two sides of the same coin. Montessori envisioned a classroom where children would be free to make appropriate educational choices, not *any* choice whatsoever. The child's freedom is necessarily limited by respect for the other students and the need for order in the classroom. In this way, a student could feel they were *part* of the whole and still retain a sense that they were *separate* and unique.

Optimal experience theory (see Rathunde and Csikszentmihalyi, 2006), and previous research on the family context of flow (see Rathunde, 1996; 2001a), adds further insight as to why a balance of freedom and discipline is important for students' flow experience. Past research has used the closely related concepts of support and challenge to study how adolescents' school experience was affected by characteristics of their family context. A supportive family context gave children freedom to be themselves, take risks, and let their interests and feelings activate the learning process. A challenging family expected children to try their best, respect the views of others, and do the work required to fully develop interests and complete tasks. Results from the study showed that the presence of *both* dimensions was associated with full attention and flow experience; whereas families that were supportive but not challenging, or challenging and not supportive, were linked with divided experiential states that split children's affective and cognitive involvement. More specifically, adolescents from families with high support and low challenge manifested a pattern of fooling around (that is, good moods with low intensity concentration on goals); whereas adolescents from families with low support and high challenge showed a drudgery pattern (that is, a high concentration on goals with low moods and energy) (see Rathunde, 1996, 2001a/b). Promoting a supportive and challenging school context (or freedom and discipline) is not a new or groundbreaking idea. However, valuable insights are gained by filtering this traditional wisdom through an experiential perspective. Research under the umbrella of optimal experience confirms Montessori insights about freedom and discipline. In other words, the most efficient use of attention and the most engaging experiences are associated with the interplay of playful/affective and serious/cognitive modes (Dewey, 1910); and such a coordination of modes is more likely to occur in contexts that support students' freedom while challenging them with discipline. Such contexts promote both affective and cognitive involvement, and both are needed to trigger flow experiences. Undisciplined children who are allowed to jump haphazardly from one interest to the next have scattered and weak concentration; conversely, children who are denied freedom and told what to concentrate on lack emotional investment in what they do. The former context is likely to be over-arousing and the latter under-arousing. For different reasons, then, children in these unbalanced contexts are less likely to experience flow.

Balancing body and mind. The two balances just discussed are important parts of Montessori philosophy. The experiential perspective that interprets each gives them a distinctly Montessori flavor. Nevertheless, balancing skills and challenges or freedom and discipline in schools is not a practice that distinguish the Montessori method from other educational approaches. A third aspect of the prepared school environment—the importance of connecting body and mind—better fits this purpose. E. M. Standing (1984, 159) noted in his biography of Montessori, "More than in any other system of education, her whole method is based on a deep understanding of the relationship between these two elements—mind and body." In this sense, Maria Montessori anticipated an important idea that is currently unfolding across many disciplines—the embodied mind. *Of the many ways in which Montessori anticipated contemporary views on child development, this orientation toward embodiment may turn out to be the most prescient and important.*

Emerging trends in modern neuroscience, cognitive psychology, and philosophy (see Damasio, 1994; Gallagher, 2005; Lakoff and Johnson, 1999; Leder, 1990) are providing new support for Montessori's suggestion that the senses, movement, the use of the hand, and other forms of embodiment, provide the foundation for the abstract processing of the mind. The notion of an embodied mind, or embodied cognition, recognizes the essential role that bodily experience plays in even the most abstract thought processes, even though this role may be relatively invisible and difficult to recognize. Many scholars are recognizing that the "higher" mind is not anchored in a realm of universal ideas; it is anchored in the physical body and the senses. For example, our concepts about space (for example, up, down, forward, back) are not, as suggested by the intellectual tradition in the West, universal concepts mapping an independently existing universe. Rather, they are abstractions made possible by our experience of the body moving in the world (see Lakoff and Johnson, 1999).

The profound shift in perspective that results from this subversion of detached reason may not be immediately apparent. It takes time to assimilate the idea that all thought carries an echo, so to speak, of the body and its sensual contact with the world. However, the implicit challenge offered to education by this new view is that every concept, no matter how abstract and presumably objective, has an experiential basis that is situated in the way the human body interacts with the world it inhabits. Even when we speak informally and metaphorically about being "an upright person" or that we have to "stay on top of things" to reach a goal, we are relying on a type of reasoning that is tied to the structure of the body as it has evolved in human history.

Montessori was an early practitioner of embodied education. She thought that the most important path to a child's natural gifts (that is, a path toward normalization) was "activity concentrated on some task that requires movement of the hands guided by the intellect" (Montessori, 1966, 138). In other words, it was the *combination* of body/senses and intellect—not one or the other—that was key to education. She comments (1973, 24–25): "It is essential for the child, in all periods of his life, to have the possibilities of activities carried out by himself in order to preserve the equilibrium between acting and thinking. . . . [otherwise] His thoughts could . . . have the tendency to lose themselves in abstraction by reasoning without end." The quotes suggest how learning is embodied, and how taking the mind out of the body leads to a number of unhealthy and negative outcomes, including excessive abstraction. Without the intellect, there would be a lack of guidance and direction; without the affectively rich processes of the body (hands-on activity, movement, emotion, and so on), there would be a lack of interest and meaning.

Many aspects of the Montessori method are tied to this general body–mind principle; one key example is the use of movement in the classroom. Children are encouraged to actively explore their environment. Everything from the layout of the classroom (see Dyck, 2002) to the design of child-sized tables and chairs is there to encourage such activity. One of the better-known Montessori techniques—the exercises of practical life (for example, table washing)—uses the hands in coordination with a goal-directed purpose. Most academic subjects, in addition, have hands-on materials to encourage activity and movement (see Montessori, 1917).

Perhaps the most recognizable way that Montessori education is rooted in the body is the use of specialized materials to educate the senses. The idea of educating the senses is unfamiliar to traditional schooling, but it provides the very foundation of Montessori education in early childhood. Montessori (1988, 148) comments, "By multiplying sense experiences and developing the ability to evaluate the smallest differences in stimuli, one's sensibilities are refined and one's pleasures increased." Through the use of a variety of materials—sound-cylinders to match like-sounds, color-tablets to discriminate subtle hues, sandpaper tablets to feel gradations from rough to smooth—young children are encouraged to refine their perceptions. Even the design of these materials invites the hand to touch. Montessori thought the materials should be beautiful and aesthetically pleasing; therefore, they were generally made with wood and glass, not cheap plastic.

In addition to generating sense experience, the didactic purpose of Montessori's sensorial activities is to develop the ability to cognitively evaluate differences among stimuli. Angeline Lillard's (2005, 65) recent book on Montessori suggests the same: "In all these exercises, movement of the body is closely entwined with cognition, since every learning exercise involves materials that children touch and move, bringing concrete embodiment to abstract concepts." Through these exercises, therefore, the child learns to use language in a more accurate way to describe the world. Lillard calls attention to recent findings on brain development in neuroscience and suggests that early experience with perceptual discrimination, particularly when the brain is young and capable of change, feeds into higher-level abilities. Early perceptual experience of the kind that Montessori education provides, in other words, affects the neural architecture of the brain, including language ability, memory, and processing efficiency.

Another essential way that Montessori brought embodiment to abstract concepts was through exposing children to nature. It is fair to say that Maria Montessori was far ahead of her time in taking the beauty of nature seriously and emphasizing its educational value. Nurturing a sense of awe and wonder through frequent contact with nature was integral to Montessori's philosophy. It continues to be a central emphasis of many

Montessori schools today where students leave the classroom to take nature walks, tend a garden, care for animals, or use nature as an extension of the classroom for study. Some of Montessori's clearest and most powerful remarks on this topic occur in her remarks about the *grandeur* of nature. She advocates appealing to the soul of a child to stimulate imagination and interest: "What he learns must be interesting, must be fascinating. *We must give him grandeur.* To begin with, let us present him with the world" (1973, 37, emphasis in original).

Montessori (1973) felt that the grandeur of nature fueled children's imagination and motivated their intellectual work. Such a view of nature is consistent with Montessori's overall linking emotion and cognition: "The child should love everything that he learns, for his mental and emotional growths are linked. Whatever is presented to him must be made beautiful and clear, striking his imagination. Once this love has been kindled, all problems confronting the educationist will disappear" (Montessori, 1989, 17). In other words, the sense of awe and mystery experienced in nature would help a student forge an emotional connection to his or her studies of the physical world.

In summary, Montessori philosophy exemplifies embodied education. An embodied education curriculum emphasizes the holistic (affective-cognitive) engagement of students and values flow in the classroom; it prepares school environments based on a sound understanding of human nature and the parameters of optimal arousal by balancing skills with challenges and freedom with discipline; finally, it provides opportunities for full human involvement, not just a disembodied intellectual focus, by valuing multifaceted ways to interconnect the body and the mind, both in the classroom and in nature.

Empirical Support for Montessori Education

To assess whether or not Montessori schools facilitated students' flow experience, a recent study in 2002 compared five Montessori schools and six traditional schools that were carefully matched in terms of social/economic status and other important school and family background variables (see Rathunde and Csikszentmihalyi, 2005a, b). The main differences between the schools were their pedagogical approaches. The school contexts in the five Montessori middle schools were in line with the embodied education perspective articulated here and many of the reforms suggested by other education researchers (see Ames, 1992; Anderman et al., 1999). For example, the Montessori schools focused on deep concentration and were imbued with a philosophy of intrinsic motivation, students had freedom to select projects (for example, students at all the schools had several hours per day for self-directed activities), students participated in school decision-making, grades were not mandatory, a significant portion of daily time was unstructured and could be used for cooperative group projects, and flexible time scheduling allowed teachers to expand or contract contact time with students depending on what was needed that particular day.

The study compared the responses of 6th and 8th grade Montessori and traditional students to the Experience Sampling Method (ESM). The ESM uses watches programmed to signal students approximately eight times per day between the hours of 7:30 am and 10:30 pm for seven consecutive days (see Csikszentmihalyi, Rathunde and Whalen, 1997). When the watches "beeped," students took out response forms and answered questions about what they were feeling at the moment, where they were at, what they were thinking about, and other questions about their momentary experience. Students in both samples also completed a detailed questionnaire about their backgrounds.

The main ESM measures used in the study were: *affect* (general mood), *potency* (energy level), *salience* (feelings that an activity was important), *intrinsic motivation* (sense of enjoyment and interest), and *flow*. Four interest quadrants were also created by combining the intrinsic motivation and salience variables: *undivided interest* (combination of the above average intrinsic motivation and salience), *disinterest* (the opposite of undivided interest, or times when intrinsic motivation and salience were *both* below average), *fooling* (above-average intrinsic motivation with below-average salience), and *drudgery* (above average salience and below average intrinsic motivation).

It was hypothesized, consistent with the argument in this chapter, that students in Montessori middle schools would report more flow and more frequent experiences combining positive affect and cognition (more undivided interest). These differences were expected to occur in academic rather than nonacademic activities. In other words, nonacademic activities were outside the mission of the school and were presumably less influenced by pedagogical differences. In nonacademic circumstances, therefore, students were expected to report similar experiences. Multivariate analysis of covariance (MANCOVA) was used for analyses.

The ESM results were based on approximately 4000 signals collected from the Montessori and traditional students while they were at school; about 2500 of those 4000 signals captured students doing academic work. Overall, the set of experiential findings showed that Montessori students reported a significantly better quality of experience in academic work than the traditional students (for a summary of the non-experiential findings not reported here, see Rathunde and Csikszentmihalyi, 2005b). There were strong differences suggesting that Montessori students were feeling more active, strong, and happy in academic work. They were also enjoying themselves more, they were more interested in what they were doing, and they wanted to be doing academic work more than the traditional students. In addition, Montessori students reported a significantly higher percentage of flow experience while working on school activities. In nonacademic work, the expectation was that when students were eating lunch, hanging out in the halls, talking with friends, and so on, that they would report similar experiences. In other words, in activities that were outside the missions of the schools, we did not expect to see many positive experiential differences in favor of Montessori students. This lack of difference is what we found across the majority of variables.

Figure 10.1 illustrates one of the more telling findings from the study—the combinations of intrinsic motivation and salience reported by the Montessori and traditional students. There are two striking differences in the figure. First, the Montessori students reported a significantly higher percentage of undivided interest. In other words, Montessori students reported above-average intrinsic motivation *and* above-average importance 40 percent of the time in academic work. In comparison, the traditional students reported undivided interest only 24 percent of the time. The primary experience for the traditional students (44 percent) was what John Dewey referred to as drudgery. Drudgery is the feeling that what one is doing is

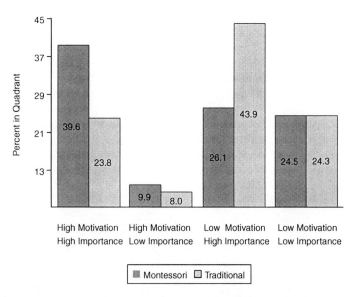

Figure 10.1 Student Experience in Montessori and Traditional Schools

important to future goals, but pursuing those goals is not motivating or enjoyable at the moment. The traditional students' high levels of drudgery serves to confirm the problem often reported in the middle school literature that traditional schools often focus on achievement or performance goals in ways that undermine intrinsic motivation and the emotional investment of students.

The findings provide important empirical support of the Montessori community. On the most basic level, the results are a confirmation that the pedagogical approach facilitates students' intrinsic motivation, flow experience, and combination of positive affect with a strong focus on academic goals. There has been very little independent research that has assessed the benefits of the Montessori approach; this is especially true in relation to middle schools and young adolescents. This study provided some confirmation that the embodied/experiential focus of Montessori education pays off in terms of student experience. The school practices were apparently in line with Maria Montessori's valuing of deeply engaging experiences: "The paths the child follows in the active 'construction' of his individuality are indeed identical with those followed by the genius. His characteristics are absorbed attention, a profound concentration which isolates him from all the stimuli of his environment" (Montessori, 1965, 218–219).

Conclusion

One of the themes in the present volume on alternative forms of education is to expand the perspective of school improvement by focusing on education that fosters holistic development and enables young people to function more effectively in modern, diverse societies. I believe Montessori education has much to offer in this regard. First, the experiential framework invoked by Montessori education and optimal experience (flow) theory offers a distinctive interpretive lens that can highlight often overlooked facets of the educational process, especially in terms of holistic development. When the quality of subjective experience and the intensity of a student's focus are not part of the story of education, positive outcomes tend to be discussed only from the "outside" in terms of measurable gains in knowledge. A greater integration of experiential ideas would complement traditional forms of instruction and provide a more well-rounded and accurate depiction of quality education. It could help provide a better framework for understanding the phenomenology of the learning and the importance of embodied education processes.

Such a framework could also help educators understand why children's motivation and concentration often falter after elementary school (Eccles, et al., 1993), and may provide insights that can counter the drift toward disembodied practices that take activity, movement, and emotion out of the learning process. Sterile educational practices are common at this point in history because of wrongheaded views of human nature and how we think and learn. Montessori's insistence on connecting doing and thinking, emotion and cognition, and body and mind is in line with emerging interdisciplinary perspectives on the embodied mind; and this convergence of perspectives may provide a useful alternative framework for thinking about preparing students for lifelong learning, rather than short-term performance on normative tests.

The main goal of Montessori education is to enhance students' focused concentration, not just for the moment or for a particular day but for a lifetime. The idea of normalization, or the normalized child, suggests that when a student is socialized in appropriate contexts (that is, proper balances of skills and challenges, freedom and discipline, body and mind, and so on), the child will be better able to recreate those moments of deep engagement on their own and stay on a path of lifelong learning. Despite the awkward term, the idea of normalization holds promise as a goal of holistic education. Optimal experience theory employs a related concept—*psychological complexity* (see Rathunde and Csikszentmihalyi, 2006). Psychological complexity refers to the self-regulative capacity to find flow; and this capacity is seen as dependent upon socialization in contexts that provide a good fit with the optimal arousal parameters of human nature. Like Montessori school contexts, a good fit is thought to occur in balanced contexts that provide freedom and discipline, novelty and order, and opportunities for differentiation and integration.

Outcomes such as normalization or psychological complexity are similar in import to the notion that schools should help students learn how to learn; in this case, however, the more accurate translation would be that schools should help students learn how to follow their interests and sense of engagement. In other words, students should be given opportunities to learn how to regulate their experience and develop their passions and talents. One of the defining characteristics of such a student would be the *undivided* and holistic nature of their attention. Such a student would not be inclined to separate their affect from their cognition, as has often been counseled in Western thought and modern, positivistic science. Such a student would not associate rational thought only with a scientist and spontaneity and passion only with an artist.

Schools that socialize normalized or psychologically complex students would have an advantage in promoting students' creativity and effective functioning in an increasingly diverse and fast-changing world. Several theories of creativity have recognized the benefits that flow from the coordination of affective-cognitive modes. Heinz Werner (1956; see also

Crain, 2000), for instance, suggested the creative person was able to combine *physiognomic perception* and *geometric-technical thinking*. The former is an immediate, body-based mode of perception—typically associated with children and artists—where a person feels emotionally connected and immersed in what they perceive. The latter mode, often linked with adult or scientific thinking, involves the use of constructs to articulate and represent our perceptions. Werner suggested that a creative person had *mobility* between these complementary modes and could, therefore, more fully engage and develop an experience. John Dewey's (1934) perspective in *Art as Experience* says much the same thing.

Research on psychological complexity has also shown that it is a worthwhile outcome to foster in schools. Talented adolescents who were able to synchronize their affective and cognitive engagement while working in their chosen domains developed their talents to a greater degree by the end of high school (Csikszentmihalyi, Rathunde, Whalen, 1993). Interviews with over ninety creative and eminent individuals (see Csikszentmihalyi, 1996; Rathunde, 1995) found numerous accounts associating creative and productive moments with the synthesis of affective and rational modes (see also Rathunde and Csikszentmihalyi, 2006). Jonas Salk, the inventor of the first successful polio vaccine, for example, described peak moments of growth and creativity as a combining intuition and rationality: "I speak of going from the intuition department to the reasoning department and then back and forth to check it out to make sure it's still true, so to speak."

Woods (2005) has argued that successful and creative schools foster the integration of students' emotional senses with their analytic capacities, a combination that allows cognition (for example, critical thinking) to be informed and inspired by felt emotion. He further suggests that this combination is indispensable for a functional democratic society. The current dominance of instrumental rationality in modern societies, often economically motivated, can be dehumanizing. Affective qualities orient individuals toward a sense of social justice and other important human values; therefore, they are essential for guiding the organizational and decision-making capabilities of analytic thought (see Woods, 2005). If it is also true, as many developmental thinkers have argued, that wisdom requires a synthesis of affective and analytic modes (see Labouvie-Vief, 1994), producing wise leaders for democratic societies will heavily depend on creating conditions in schools that foster the integration of these complementary human capacities.

In a complex world where challenges grow at the pace of globalization, schools need to foster students' psychological complexity, self-regulation, engagement, creativity, and wisdom in addition to their competence in basic domains of knowledge. Montessori education is not the only way to accomplish these lofty goals; nevertheless, it would be worthwhile when considering alternative forms of education to take account of the embodied practices of Montessori schools.

Note

1. The historian Natalie Davis beautifully described such a state during an interview for a creativity study. In her peak creative moments she feels as if there is a side to her "that is absolutely carried away, floating along with the project; and a side that is also detached and looking at myself." The former provides affective intensity and the latter the ability for objectivity and criticism (see Rathunde, 1995).

Chapter 11

Education for Freedom: The Goal of Steiner/Waldorf Schools

Martin Ashley

What is Steiner/Waldorf Education?

To understand Steiner/Waldorf education, it is necessary to journey back to the Germany of 1919 that stood in social ruin at the end of the Great War. Thinking-people were in despair at the ravages of social inequality compounded by national defeat. It was a time receptive to radicalism. These were the conditions that allowed Rudolf Steiner to present his ideas for a new social order, based upon a radical reinterpretation of the time-honored notion of liberty, equality and fraternity. Separation of the three spheres of culture (freedom), rights (equality), and economics (fraternity), a principle known as social threefolding, was fundamental to this. Underlying Steiner's entire philosophy was the primacy of freedom. Education comes into the sphere of culture, and it is absolutely fundamental that the school should serve the child, not the state. Steiner/Waldorf schools have, as their ultimate goal, the development of fully free human beings, but they operate from the postulate that freedom does not exist simply by virtue of an arbitrary declaration of human rights. For Steiner/Waldorf schools, freedom cannot be a method of education but must be the end result of it (Carlgren, 1972).

Steiner's output of scientific and philosophical writing was immense. Much of it is highly esoteric and grounded in a spiritual-scientific system of thought known as anthroposophy. Anthroposophy was a word developed by Steiner after breaking away from earlier roots in theosophy, and it literally means "wisdom of man." Steiner claimed that the mystical operation of the mind could and should transcend natural science and that such powers would be available to anyone prepared to cultivate the necessary mental faculties (Rawson and Richter, 2000). At the very core of anthroposophy lies the deep-rooted philosophy of freedom that seeks to liberate the human condition through the integration of science, art, and religion. Steiner's use of the term has much to do with the freedom to form concepts out of the reality in which we live as human beings—an ultimate form of individual sovereignty. It is to do with our ability to attain knowledge outside the bounds set artificially by an empiricism limited by an ultimate, unchangeable, "objective" and "material" reality (Nordwall, 1980).

Steiner founded his first school in 1919 to serve the children of employees at the Waldorf-Astoria cigarette factory in Stuttgart, hence the sometimes encountered term "Waldorf" as an alternative to "Steiner" school. He had been invited to do so by the factory owner, Emil Molt, and the workers were captivated by the imagination and vision of a radical, new form of education in which the children of workers and management would be educated together throughout a full twelve grades (ages 6–18). The Steiner/Waldorf schools are now a worldwide movement that constitutes one of the most strongly established alternatives to state schooling. There are 870 schools globally in sixty countries including most European countries, Australia, Canada, Egypt, India, Israel, Japan, Kenya, New Zealand, South Africa, South America, and the United States (Woods, P. A., et al., 2005). Twenty-three of these schools are in England and include large, well-established schools catering for the entire 5–18+ age range and small schools of relatively recent foundation employing only two or three full-time teachers.

Steiner/Waldorf schools have always been coeducational, and Steiner emphasized that the education should be available to all, which creates a tension in that the schools in England are all, at the time of writing, independent. This means that parents must pay fees, although these are generally lower than those paid in other private schools and sometimes deferred through donations in kind or through a sliding scale based on need or income. There have been protracted and difficult negotiations to obtain state funding for one of the schools under the UK government's academy schools program, which aims to invite sponsorship of state schools by private capital in order to increase diversity and choice.

Anthroposophy underpins a pedagogy of education toward freedom that sees schools and teachers charged with the sacred task of helping the child's threefold being (body, soul, and spirit) to incarnate. The role of schools and teachers in helping a child to incarnate in the world out of a previous spiritual existence is a tenet that would be problematic for many secular teachers (and parents) in state schools. A detailed and ongoing study of child development is seen as absolutely fundamental to the work of Steiner/Waldorf schools. The curriculum has been developed to follow closely the way in which the interests and aptitudes of children change with growth. Steiner/Waldorf teachers receive in-depth education about child development and, importantly, continue the development of this understanding throughout their careers through the ongoing child study that takes place in the schools. Steiner/Waldorf teachers are expected to engage in "inner work." The nearest equivalent in state schooling might be "reflective teaching," but "inner work" requires daily meditative practice, often focused upon a particular child, or on a class's needs.

In practical terms the teachers' task is to facilitate children's growth through a series of stages that posit the immature child as qualitatively different in various ways to the mature adult. Crucially, a distinction is made between the dominance of *willing* (control of limbs) during the 0–7 years, *feeling* (an intense aesthetic sense) during the 7–14 years, and *thinking* (the unfolding of rationality) during the 14–21 years. The second dentition (emergence of the adult teeth) is the principal outward sign that the child is ready to pass on to the aesthetic phase of education. It is at puberty that the young person comes to make judgments, form independent concepts, gradually direct behavior according to conscious intentions motivated by ideals and thus pass to the thinking phase.

Until the age of fourteen, children are under the direct control of a class teacher with whom they remain, ideally, for eight years. On progression to upper school at age fourteen, a class guardian assumes responsibility for the close oversight of the young people. The quasi-familial development of each class cohort as a social group who know each other and are known quite deeply is one of the most significant distinguishing features of Steiner/Waldorf education. Small but important rituals, such as students shaking hands individually with the teacher at the beginning of each session, are visible signs of this. Much faith is invested in the class teacher as the *authority who represents the world to the child*. More will be said later about the significance of helping the child grow toward freedom through authority. The autonomous, collegiate, professional status of the teachers is jealously guarded and Steiner/Waldorf schools would be highly resistant to centralist prescription of pedagogy.

Much, in turn, is expected of the teachers. During the early years, the child must learn that the world is a good place. The kindergarten teacher must model this for the children through her own behavior. Steiner stressed that the actual words spoken to the child are relatively unimportant. It is the teacher's mood and actions that reverberate in the child's mind. During the middle years, the child must learn that the world is a beautiful place. A Steiner/Waldorf teacher needs quite a degree of personal artistic accomplishment. He must be able, for example, to produce beautiful blackboard drawings. Above all, he must be a story teller. He must not hide behind technology such as books or computers but must be able to bring things directly to the children through his own voice, through speaking and through singing. During the upper school years, the teacher must expect to be challenged by the students who will test out her knowledge and understanding of the subject. Steiner indicated that it is natural for teenagers to do this and that the teacher must have enough subject knowledge to maintain a credible position as an *authority on the subject.*

To understand Steiner/Waldorf education, it is essential to understand just how much importance is attached to the process of human interaction in facilitating learning and growth. It is true that computers are very unlikely to be found in classrooms. However, computer suites are found in upper schools, computers are used in school administration, and Steiner/Waldorf-educated students can be perfectly proficient in information and communication technology (ICT) skills on leaving school. ICT is absent from classes below upper school because of the degree to which Steiner emphasized that children learn primarily through human contact. Even books, in this sense, are a "technology" that can stand in the way. Children do use books, but as a resource to turn to after inspiration by the teacher. Steiner/Waldorf schools are often quite strongly aligned with movements orientated to the protection of childhood, and teachers are seen as necessary to mediate for children in a world saturated by information but lacking wisdom.

The portrait of a stern Steiner, found almost inevitably in staffrooms and entrance halls of Steiner/Waldorf schools, can perhaps convey to the uninitiated visitor the impression of cult status. Such an impression would not be entirely without foundation, as the schools continue to look to their founder and strive to follow his teachings. Steiner set out his educational principles initially in 1919 in a course entitled *Education for the People* and subsequently in an ongoing series of lectures and "pedagogical meetings," known as the *Konferenzen,* to the teachers who were to create the new schools. There are now real tensions around maintaining the integrity of the Steiner philosophy and openness to other ideas. An important study by Masters (1996) evaluated the degree to which present-day teachers are in fact faithful to the original *Konferenzen* principles, with predictably mixed results. Nevertheless outside observers often note that Steiner/Waldorf educators are very clear about what they believe and are aiming to achieve. Steiner/Waldorf education has evolved slowly, the principles of its founder acting as something of an anchor against extreme fashion swings. Teachers face growing tensions between hard-line conservatism and the need to compromise with regard to seemingly unstoppable developments in mainstream education, such as the proliferation of new technologies or innovation in accountability and assessment practices.

This is reflected in the research that has been undertaken into Steiner/Waldorf education. Any bibliography on Steiner/Waldorf education will be heavily dominated by the writings of Steiner himself, and the concern of other writers within the Steiner/Waldorf movement has been primarily with the interpretation of these writings.

The Research Study

This section of the chapter draws principally upon the research carried out in England in 2004–2005 for the UK government's education department (DfES)[1] by the author and two colleagues, reported in detail in Woods, P. A., et al., (2005), in which twenty-one of the twenty-three Steiner/Waldorf schools in England participated. The study involved questionnaire surveys, fieldwork visits, and case studies of good practice. This research, being carried out at a "critical distance" by independent researchers, marked a distinct departure from the tradition of research within the Steiner/Waldorf movement. Generally, it has been well received within the movement.

I shall first outline seven broad themes that arose out of this work, drawing also on the literature that was surveyed, which is detailed in our report (Woods, P. A., et al., 2005). The first two are particularly related to the holistic approach of the schools. I shall then discuss briefly some significant issues relating to Steiner/Waldorf education.

Seven Broad Themes

Working from the Whole to the Parts
Rawson and Richter (2000) argue that the whole structure of the Steiner/Waldorf curriculum is profoundly ecological and preceded present-day concerns for sustainable development through ecological awareness by several decades. Teaching methods always work from the whole to the parts and all phenomena are taught as though they had an animating, life principle or being. It is thought that initially an emotional connection to any phenomenon must be awakened in the child. Conversely, a reductionist, analytical approach that attempts to synthesize meaning from fragments of knowledge without first presenting a bigger picture that is coherent to the child is avoided.

Attentiveness to Child Development
It was the grounding in child development that respondents identified as the most distinctive feature of Steiner/Waldorf education in the survey of Steiner/Waldorf Schools in England. All of the twenty-one schools in the research fully agreed that a grounding in child development was a distinguishing feature of Steiner/Waldorf education, the only such attribute achieving such a unanimous level of support (Woods, P. A., et al., 2005, 66). Given the degree to which the understanding of child development

has been abandoned or downplayed in state teacher training in the UK, the close attention to how children grow and develop might be seen as a particular strength, at least by those who hold that it *is* important.

Numerous examples are found in the *Steiner Schools in England* report of how a particular belief about the nature of children at a given age affects the curriculum. For example, the Romans are taught for the first time to Class 6 (aged 12) because of the link between law, confrontation, rationality and the more confrontational disposition and emerging rationality of children of that age. The emphasis on food and farming, building and outdoor work in Class 3 would be another particularly significant example, given the belief that children at this age (9 years) "cross the Rubicon" or undergo a psychological crisis in which the world becomes much more of an external reality. Activities that show the child how humankind makes the world a good place to be in are therefore considered appropriate.

Aiming for Freedom

The Steiner/Waldorf curriculum does not progressively narrow at key transition points such as ages 14 and 16. Students continue to study all the subjects until the end of sixth form (ages 17 to 18), and this may contribute to the articulacy of the young people as well as their ability to take up options at later dates. Ogeltree (1998) in a large international study of 234 Steiner/Waldorf schools found that over 90 percent of the teachers believed that Steiner/Waldorf education develops free-thinking individuals. Smith (1998) found that two thirds of 150 questionnaires returned by Steiner/Waldorf schools, training centers, colleges, and conference participants in the United States and Canada supported the proposition that Steiner/Waldorf education develops a strong sense of self and good life skills. She also found, however, that less than half of her respondents felt that it encouraged young people to develop a sense of responsibility to the wider community.

Given that Steiner/Waldorf education is fundamentally about the development of free-thinking individuals, it is reasonable to ask whether this is achieved. It is difficult, however, to assess whether Steiner/Waldorf schools really do achieve this aim. We were presented with much anecdotal evidence during our own study, but as with the two studies above, it was compromised through originating within the Steiner/Waldorf movement itself. An example would be the school that claimed their students did very well in a wide range of subjects studied at a level of advanced specialism for 17–18-year-olds (the English "A" level) without having taken the public examinations at 16+ in the same subjects. This school offered only three public exam subjects at the latter level but claimed that future success was due to superior student maturity and thinking ability.

Relationships with Students

The eight year relationship of the class teacher with 7–14-year-olds is possibly the most distinctive single feature of Steiner/Waldorf pedagogical practice, and it naturally raises questions in the minds of teachers in other sectors. It is important to state that the class teacher does not spend the whole day with the students. It is the two hour "main lesson" at the beginning of each day that is the main focus of this relationship. After main lesson, there are four to five subject lessons. Some of these will be taken by the class teachers, others by specialist teachers. Early on in our study, a respondent suggested that there were three types of teachers in Steiner/Waldorf schools:

- a class teacher teaching his or her own class,
- a class teacher teaching his or her subject to another class,
- a subject teacher who is not a class teacher.

This respondent suggested that there was a continuum of diminishing quality in class control and teacher/student relationship from the first to the last of these. This proposition was put to subsequent interviewees, all of whom endorsed it as concordant with their own experiences. The weight of our observational evidence supported this quite strongly. Students were more settled and generally behaved better with their own class teachers even (with one exception) up to the age of fourteen, and the uninterrupted focus on learning often outweighed any benefits of specialists' greater subject knowledge (Ashley, 2005).

In our survey of schools, 91 percent (n = 21) saw "other teachers being available for support/consultation" as "very important" in addressing potential deficiencies in a class teacher's subject knowledge, whereas only 28 percent saw specialists taking the main lesson as very important (Woods, P. A., et al., 2005, 73). The balance of evidence would seem to suggest that the eight-year relationship works and is beneficial to student development. It is concordant with the notion that spiritual growth and holistic development through close and caring relationships are a positive characteristic of the Steiner/Waldorf schools (Petrash, 2003). The overall suggestion is that care rather than discipline is at the heart of the schools' functioning.

McDermott et al. (1996) found that the Steiner/Waldorf-inspired public school in a deprived and challenging district of Milwaukee achieved better grade results with African-American students than other schools in the district, attributing this primarily to more stable student–teacher relationships, a finding replicated by Byers et al. (1996). Schieffer and Busse (2001) claim, cautiously, to have identified similar results in a small quantitative study. Rivers and Soutter (1996) have investigated bullying and found that, whilst there is teasing, classes "look after their own" and are less likely to victimize unpopular children. Easton (1997) claims, as a result of interviews with more than fifty students (grades 7 to 12, from three Steiner/Waldorf schools in the United States), that the caring, communitarian ethos of the schools is appreciated by the students. House (2001), however, found that inefficient management practices associated with the collegiate system did not necessarily result in this well-being extending to all teachers, a finding that echoes some of our own conclusions, which will be discussed shortly.

The Oral Tradition
A key finding of the *Steiner Schools in England* study was the strength of the oral tradition in the Steiner/Waldorf schools. Ward (2001) confirms the study's observations that Steiner/Waldorf teachers are effective storytellers who operate within a much stronger oral tradition than is the case in state schools. Vocal health and the ability to articulate clearly are given high priority for teachers and students. Recitation and choral speaking feature regularly in the curriculum. This is manifest in the teaching of modern foreign languages as well as permeating most lessons taught in English. In most schools, two modern European languages would be taught to children of primary age upward and the European ethos of the Steiner/Waldorf schools is strong. We observed not only the teaching of songs but sometimes quite extended dramas in which the children spoke the parts in French or German.

The Steiner/Waldorf approach stresses the importance of building all literacy on the foundations of oral skills or the "living word," and the schools are resistant to criticism that the formal teaching of reading and writing begins too late.[2] Coupled with the stress on rhythm, movement, and drawing, there are grounds to suggest that boys lagging behind girls in fine motor skills might fare better under the Steiner/Waldorf approach with its close attentiveness to the characteristics of children—for example, the way in which grammar teaching starts with the verb because it is believed that the "action word" most closely corresponds to the natural movement of the young child. A large-scale investigation of this important possibility would be strongly justified.

Rituals and Routines
In the *Steiner Schools in England* study we observed much ritual, not only in classrooms but in teachers' meetings too. Each day is marked out by each class through the rituals of the opening and closing verse. Oberman (1997) highlights the role of ritual, routine, and ceremony in maintaining the continuity and ethos of the Steiner/Waldorf schools, and there is little doubt that, from the attention that is given to regular large-scale festivals right down to

the protracted procedures of a kindergarten meal, ritual is significantly more visible in Steiner/Waldorf than in state schooling. Henry (1992) regards such display favorably as evidence of a communitarian and social rather than individualist ethos. In a comparative ethnographic study, she comments on such matters as how an egalitarian, nature-orientated community metaphor contrasts with the "rule by time and efficiency" of a private college prep school. Astley and Jackson (2000), however, refer critically to this study in questioning the belief that kindergarten children are pre-rational and suggest that they create their own meanings for rituals, held in abeyance, but different to the anthroposophical meanings of the teachers.

Art and Creativity
Second only to the misconception that Steiner/Waldorf schools are free schools in the tradition of A. S. Neill's *Summerhill*, the misconception that was most frequently encountered by Steiner/Waldorf teachers amongst the non-Steiner/Waldorf community was that the Steiner/Waldorf schools teach predominantly an art curriculum because there is a lack of rigor with regard to other subjects. The idea that there is a lack of rigor or a narrow concentration on art is not supported by the *Steiner Schools in England* study. Art, however, was fundamental to the egalitarian principles of the first Steiner/Waldorf school. Steiner wrote of the alienation of the worker from the cultural traditions that allowed him to take sustenance from spiritual sources (1919, 3), astutely observing how the higher classes had avoided this fate through appropriating art to themselves. It is necessary also to reiterate that the "art" is there because it occupies a fundamental place in all pedagogy. For example, it is believed that in early literacy, pictures precede letters. Phonic symbols emerge out of a combination of drawing, recitation, rhythmic movement, and fairy tale. The Steiner/Waldorf early years teacher does not use the ready-made symbols of a commercial scheme such as *Jolly Phonics* (Lloyd and Wernham, 2005), but she must create the symbols out of her own artistry and closeness to her children's learning.

Much of the art instruction observed in our study was actually very formal. Children aged between 8 and 11 were often seen copying techniques and ideas directly from the teacher. There are pedagogical reasons for this and it would certainly not be the case that children in Steiner/Waldorf schools are just free to dabble in self-expression all day. There is undoubtedly a tension between Steiner's views that children go through stages, such as a copying stage of development, and the currently enthusiastic reception that has been given to such ideas as Gardner's multiple intelligence or the notion of catering for individual learning styles. Whilst there has been a tendency in some state discourse to talk of "catering for the needs of the visual learners," Steiner would maintain that all children go through an important visual stage, linked to the idea that mankind itself evolved through a visual stage.

The appreciation of form through drawing is thus considered to be of fundamental importance. A class of eight-year-olds forming a moving figure of eight during a eurythmy lesson[3] would be an example of learning through the kinesthetic approach and, whilst art and the feelings dominate during the middle years, Steiner/Waldorf education would not label children as "visual learners" or "kinesthetic learners." Ogeltree (2000) set out to compare the creative thinking development of Steiner/Waldorf and state schooling using a robust, quantitative approach based on the Torrance Test of Creative Thinking Ability. He concluded that Steiner/Waldorf students had significantly higher creativity scores than their state school peers. However, he had some difficulty in explaining why the verbal fluency of students in English primary state schools was as high as students in Steiner/Waldorf schools in England (whereas it was higher amongst Steiner/Waldorf school students compared with state school students in Scotland and Germany). Cox and Rolands (2000) found robust evidence that the focus in Steiner/Waldorf schools on art as a pedagogical medium resulted in significantly more accurate observation skills as well as more mature artwork than a comparable group of state school students.

A Goethe-Based Approach to Science

It is the science curriculum that provides one of the most significant insights into the holistic, child-development-based approach to art and creativity. Steiner was heavily influenced by Goethe who espoused an alternative scientific world view. Child development principles also privilege nature and the life sciences in the primary years where there is a very significant difference from the "top down" National Curriculum for science in England in which physics and chemistry continue to be presented in such simplified forms as "a force is a push or a pull" in the earliest years of Key Stage One (5–6-year-olds).

Formal teaching of physics and chemistry does not begin until class 6 (age 11–12) for physics and class 7 for chemistry. In physics, there is a strong emphasis on observation, which might begin with the study of acoustics arising out of the idea of musical instruments more familiar to the children. This would progress to a study of the larynx in class 8 life sciences. The notion that the physical sciences are characterized by quantification in the form of measurement is disputed by the theory underpinning Steiner/Waldorf education. The "being" or essential nature of the phenomena to be studied is thought to be lost through a purely quantitative approach. The qualitative changes in children's thinking are reflected in the view that the fairy tales of class 1 (age 6–7) contain the imaginative pictures that will support the kind of paradoxical thinking demanded by an appreciation of quantum theory in class 12 (age 17–18).

Conventional reductionist models such as the Niels Bohr model of the atom that is still found in many conventional school text books are avoided on the grounds that they confuse children's thinking and actually promote misconceptions. This approach to science was evaluated in a significant, large scale study by Jelinek and Sun (2003). These authors were forced to conclude that the scientific reasoning of Steiner/Waldorf students was superior to that of comparable students in public (that is, state) schools. The Steiner/Waldorf students generally performed better on an international (TIMMS) test of scientific understanding based on magnetism. The Jelinek and Sun study analyzed many of the pedagogical principles of Steiner in some depth, with some positive findings. In particular, the study appeared to endorse the claim that students taught less content and subjected to less examination pressure did better in the long run and that Steiner/Waldorf education was successful in its aim to "educate human beings."

In spite of this, Jelinek and Sun were not convinced that Steiner/Waldorf offers a viable form of science education. Their main reservations were that some of the links between science, religion, and philosophy were inaccurate and they were not convinced by the stress placed on Goethe, still less by Steiner's claim to "supersensible knowledge." In spite of their positive findings, indeed, they concluded that "as a first step Waldorf should disregard Rudolf Steiner and anthroposophy as the source of accurate scientific concepts." There are unresolved conflicts here, principally between a science education based on "inaccurate science" that leads to better scientific understanding.

Some Significant Issues

Limitations of Developmentalism
Some might see Steiner's child developmental approach an admirable basis for a pedagogy in harmony with the child, the degree to which the curriculum follows this being seen also as a great strength. Others, however, might question that Steiner took this too far, perhaps suggesting the approach is too deterministic, placing limits on the expectations of what children might achieve. Developmentalism in such forms as Piaget's well-known stage theories has been the subject of significant and ongoing critique. Whilst Piaget is recognized as having made a huge and lasting contribution to child development, Donaldson's (1978) seminal discussion has rendered his overly rigid structure of stages increasingly doubtful as a basis for pedagogical practice (Long, 2000).

Piaget-style stage theories such as Kohlberg's model of moral development (Gilligan, 1982) or James Fowler's "faith stages" (Dillon, 2000) have faired similarly badly, being seen by feminist critics as providing a picture of the child's world distorted by a particular systemic view that is oppressive. Most often, the complaints are of children being held back because in spite of a question asked or an aptitude shown, they are considered not ready for the next stage. Mills (2005), for example, feels strongly about this. It is difficult to see why Steiner's own view of development should necessarily escape such critique and outsiders who look into anthroposophy can express alarm at the degree to which Steiner's esoteric evolutionary philosophy of cultural epochs exceeds the bounds of accepted history, philosophy, science, or theology.

Occult Status and Accusations of Racism
Some of the most sustained criticism has come from an American organization calling itself PLANS (People for Legal and Nonsectarian Schools), which has set up a Web site devoted to criticism of the Steiner/Waldorf approach. Two concerns above all others preoccupy PLANS. The first is that Steiner/Waldorf education is "occult" and the second that it is racist. I have already referred to the Steiner portraits that can give the impression of cult status. Beyond that, however, our own research found very little to substantiate the claims of PLANS with regard to the English Steiner/Waldorf schools. Conversations with students tended to validate the claims of teachers that whilst they worked "out of anthroposophy" there was no intention to convert the students to this belief system. Moreover, the current vulnerability of state schools to the hyped up marketing of pseudoscientific fads such as "brain gym" has in any case caused growing alarm in the neuroscience community (Goswami, 2006). As Goldacre (2006) recently pointed out, state schools can readily and uncritically embrace supposedly educational practices based on bad or "pseudo" science.

Inclusion
Although Steiner/Waldorf schools were amongst the first to embrace the concept of coeducation, the tendency to refer only to Steiner's own teachings has arguably lost them this advantage, as understanding and practice has developed elsewhere. Golden's (1997) three year ethnographic study of a class claims to demonstrate the degree to which patriarchy is embedded and male values woven into the stories that constitute much of the curriculum. This would be a feminist critique of the stories and legends that are so important in the curriculum, and it may well have some validity. A preliminary study by Smith (1998) indicated that a sense of responsibility to the wider community was the principal weakness of Steiner/Waldorf schools in the United States. However, this would seem not to be in keeping with Steiner's original ideas of inclusion.

Funding and Admissions

In the UK at least, problems of class and ethnicity representation may be due to the schools' alternative status. It is predominantly white parents of a certain social class and philosophical outlook who select, and therefore populate, these schools. Steiner wished, through social threefolding, to keep education and politics separate; but if the state is to pay for education, it is difficult to see how this is entirely possible unless the state itself accepts social threefolding. A related weakness that was immediately apparent in the *Steiner Schools in England* study concerned admissions and the right, claimed by teachers, to veto which new children can be admitted to a class. This is simply too far from government policy on admissions. Given that the state has assumed the responsibility of protecting the right of every child to an education, an impossible situation is created.

Social threefolding holds that, through separation of state, economy, and culture, all families, not just rich ones, should have freedom of choice in education and access to nongovernment schools for their children. Such a principle would also hold in many philosophies outside social threefolding. There is, however, an inherent conflict between the right and de facto freedom of families to choose schools and the responsibility of the state to safeguard the children of families unable or unwilling to exercise this right. An unwillingness to admit "unsuitable" children implies a lack of reciprocity that confirms Smith's (1998) finding of weak responsibility to the wider community.

The Collegiate System

It is almost inevitable that when the foundations for action are principally the teachings of a past leader, some polarization between fundamentalist, progressive, and pragmatic positions will take place. The encroachment of state funding (which is a possibility in England) seems increasingly likely to be the catalyst for such division. Steiner/Waldorf schools have no designated head teacher, being managed collectively by the teachers acting as a college. This system has some advantages, particularly in terms of teacher ownership of practice (see Woods and Woods, 2006a, b), but our study fielded frequent complaints about inefficiency, delay in decision-making, and increased stress in teachers. The lack of a clear authority figure who will give an immediate response to typical parental complaints is clearly a major issue for the parents who air their grievances on the PLANS Web site.

On this issue, there has been compromise in the appointment, at the government's behest, of a "principal" for the UK school that is proposed for state funding. Other schools have addressed the issue through the election of a "teacher manager," the delegation of significant responsibilities to an administrator, or the introduction of a mandate system that disperses decision-making responsibilities (Woods, P. A., et al., 2005, Woods and Woods, 2006b). On the vexed issue of repeated, high stakes summative testing (Standard assessment tests) with published league tables, it is the UK government that has taken the fundamentalist stance. It was made clear to us in our study that this one issue above all others was entirely non-negotiable, a principle the UK government has hitherto held to for English schools in spite of growing critique from its own supporters.

Information and Communications Technology
Fundamentalist doctrinal fervor and unwillingness to compromise can therefore hardly be seen as a weakness peculiar to the Steiner/Waldorf schools. It remains to be seen how a fundamentalist position toward information technology will play out as the technology reaches maturity. There is undoubtedly much wisdom and good sense in such Steiner/Waldorf principles as the need for teacher mediation in the face of information overload. The strength of the UK government's support for and insistence on ICT in state schools could also be seen as tending toward the fundamentalist. However, the possibility that Steiner/Waldorf children could be excluded from the evolution of new multimedia literacies that are genuinely in tune with childhood and its needs might also be seen as a fundamentalist tendency if there is an unwillingness even to evaluate new literacies in the light of established principles.

Possible Lessons and Implications

Enhancing Democracy Through Diversity

It is difficult to assess the real potential of Steiner/Waldorf education to enhance democracy through diversity. The stress placed on developing free-thinking individuals has featured several times in this chapter, but free-thinking individuals do not necessarily constitute a democracy. Uncertainty about the collegial system serves as a warning here, particularly if it is a reflective microcosm of the inefficiency and uncertainty that might pertain in an entire state run on such principles. There has also been the suggestion that the Steiner/Waldorf schools and their alumni are too indifferent toward the wider community and it is uncertain that they would address the problems of a society suffering from a surfeit of individualism.

On the other hand, the existence of significant numbers of free-thinking citizens is undoubtedly a curb on the misuse of power by the state. This possibility arises at a time when there is growing concern over the surveillance society and the state's management of terrorism, a problem that some would see as being of the state's own making in the first place. If democracy is conceived in such straightforward terms as majority voting, then some would argue that there is little real democracy in state management of foreign affairs. The really challenging question is that of whether Steiner/Waldorf education can, or logically should be, fully integrated within existing political economies, or whether the ultimate question is that of an all-or-nothing adaptation of social threefolding.

Enlarging the Purpose of Education

If the rhetoric of the National Curriculum in England is to be believed, it cannot really be claimed that the Steiner/Waldorf system enlarges the purpose of education. Nearly all its aspirations are also to be found in the rhetoric of state education. Possibly it is the very term "National" that is questionable. Though it has been criticized for its Eurocentrism, the Steiner/Waldorf curriculum is not a "national" one and challenges us to consider why any state should still want a curriculum geared to its own economic and political goals. The question, therefore, would be that of whether the Steiner/Waldorf system inherently stands any more chance of realizing those overarching ideals. The answer I would give is that if there is any lesson for state education that is difficult to contest, it has to be that of *stability*.

Steiner/Waldorf education offers stability for children through the eight-year class teacher system and the "class guardian" for upper school students. This does seem a better alternative than the fragmentation of both curriculum and pastoral support that often characterizes state practice. The two-hour "main lesson" also offers a well-refined system of curriculum integration that incorporates many good principles of "rhythm" or pace and timing. Parallels can be found here with the much more recent work of Claxton (1997), which has drawn attention to important mental processes that operate beneath the immediate level of conscious processing. In ways such as this, Steiner/Waldorf education offers stability for teachers precisely because social threefolding keeps it immune from endless government initiatives and the economic predation of private "consultants." Paradoxically then, I would thus see Steiner/Waldorf education as potentially enlarging the purpose of education not through offering anything new, but through offering the stability to achieve in reality ideals and objectives that have been around for over a century.

Lessons for Steiner/Waldorf Teachers

Some lessons, such as the possible need for a "principal," are at least being trialed at a proposed, state funded Steiner/Waldorf academy school. There will need to be an evaluation of whether the collegial system can be made more efficient and capable through some injection of revised thinking on leadership and management. A willingness to consider initiatives such as the mandate system has been shown to improve the efficiency of collegial management at a large Steiner/Waldorf School (Woods and Woods, 2006b). Another successful initiative has been developed out of the work on associative leadership by Wolf-Phillips (2006). These have shown that the lesson for alternative educators does not necessarily have to be that practices from the state system should be imported. A willingness to relinquish strict dogma and reconsider existing approaches may be equally or more effective.

In the *Steiner Schools in England* research we found unease amongst some teachers about accountability and assessment. A minority of teachers worried that state schools were well ahead in practices of assessment and recording, perhaps as a result of the publicity given to the work of Wiliam and Black (2006). The issue of accountability is at one level associated with management and leadership. At another level, it is associated with the loss of professional autonomy. Forced conformity to assessment practices tied up with summative testing and league tables is one of the most significant ways in which state teachers have lost professional autonomy.

There is undoubtedly scope for Steiner/Waldorf teachers to learn from the state sector, and for state teachers to learn from the Steiner/Waldorf sector. In the *Steiner Schools in England* report we suggest the major themes of knowledge exchange that might be pursued (Woods, P. A., et al., 2005, 122). That these are themes well worth pursuing was a fact readily acknowledged at a conference on "building bridges" held as an outcome of the research and which, for the first time, brought members of the Steiner/Waldorf and state education communities together in such a forum (Woods, G. J., et al., 2005).

Notes

1. Department for Education and Skills, the UK government education ministry for England at the time of the research, now renamed Department for Children, Schools and Families (DCSF).
2. The commencement of reading would be at around the age of six or seven and coincides with the time children start to gain their adult teeth, which is seen as a sign of developmental readiness.
3. Eurythmy is an art of movement originated and developed by Rudolf Steiner, which is meant to help children develop harmoniously with mind, body, and soul.

Further Sources of Information

Rawson, M. and Richter, T. (2000) *The Educational Tasks and Content of the Steiner Waldorf Curriculum.* Forest Row: Steiner Waldorf Schools Fellowship.

Steiner, R. (1907) *The Education of the Child*. London: Steiner Books.
Steiner, R. (1919) *The Philosophy of Freedom*. Forest Row: Rudolph Steiner Press.
Steiner, R. (1923) *The Child's Changing Consciousness as the Basis of Pedagogical Practice*. East Grinstead: Anthroposophic Press.
Woods, G. J. and Woods, P.A. (2008) "Democracy and Spiritual Awareness: Interconnections and Implications for Educational Leadership." *International Journal of Children's Spirituality* 13/2: 101–116.
Woods, G. J., Woods, P. A., and Ashley, M. (2005) *Building Bridges Conference: Summary of Outcomes – Towards a Wider Sense of Community?* Bristol: Faculty of Education, University of the West of England.
Woods, P. A., Ashley, M., and Woods, G. J. (2005) *Steiner Schools in England*. London: Department for Education and Skills (Ref No: RR645, available at www.dfes.gov.uk/research).
Woods, P. A. and Woods, G. J., (2006) *Feedback Report (Meadow Steiner School): Collegial Leadership in Action*. Aberdeen: University of Aberdeen.
Woods, P. A. and Woods, G. J., (2006) *Feedback Report (Michael Hall Steiner School): Collegial Leadership in Action*. Aberdeen: University of Aberdeen.
Woods, P. A. and Woods, G. J. (2006) "In Harmony with the Child: The Steiner Teacher as Co-leader in a Pedagogical Community." *Forum* 48/3: 217–325.
Web sources: Steiner Waldorf Schools Fellowship, http://www.steinerwaldorf.org.uk/
People for Legal and Nonsectarian Schools (PLANS), http://www.waldorfcritics.org/

Chapter 12

Pathways to Learning: Deepening Reflective Practice to Explore Democracy, Connectedness, and Spirituality

Philip A. Woods and Glenys J. Woods

In this chapter, we discuss some of the issues that arise from a consideration of the alternatives featured in the book. Specifically, the chapter

- addresses the positioning of alternatives, the different meanings and implications that the concept of alternatives can have, and questions of power bound up in it;
- articulates the value of diversity and difference in school systems, and the importance of cooperative practice; and
- sets out a framework of features, formulated as a result of our review of the alternatives, which characterizes educational settings that have a developmentally democratic character and strive to be inclusive, open to alternatives, and capable of nurturing rounded physical, spiritual, emotional, cognitive, ethical, and social development in students.

Suggestions for reflective activities, designed for both alternative and mainstream educators, are given in boxes at the end of each of the chapter's three main sections. These are intended to encourage openness to change and, specifically, the connectedness that we identify as a preeminent theme of the book and which has an inner spiritual quality as well as an outward, active aspect that transcends boundaries and challenges unjust power relations. The suggested reflective activities are based on the concepts and framework of features discussed in this chapter. They are broad pointers to areas of exploration, which educators and others may follow in their own way, selecting from and utilizing, critiquing, and adapting concepts, practices, and experiences of the alternatives discussed in previous chapters. The activities may be followed informally or formally as a part of systematic study; may be undertaken by an individual or small group, or used as a basis for a whole-school or whole-district process of reflection; and be approached at a general level or made more detailed.

Alternatives: Positioning, Power, and Meanings

The focus of this book is on what we have termed alternative education. However, the meaning of the concept of alternative cannot be taken for granted. Raymond Williams (2005, 40), for example, draws a distinction between alternative and oppositional forms of cultural and social life, arguing that there is "clearly something that we can call alternative to the effective dominant culture, and there is something else that we can call oppositional, in a true sense." If oppositional means interrupting the social reproduction of the neoliberal agenda, it involves challenging explicitly and outwardly how education is being reconfigured in contemporary times.

The alternatives in this volume express in different ways challenging visions of humankind's potentiality and of how society can and should be changed. They are characterized principally by value-rationality—in which social action is predominantly orientated to ideals that arise from "self-conscious formulation" of ultimate values (Weber, 1978, 25), as distinct from instrumental rationality where social action is dominated by a calculative spirit aimed at maximizing goal achievement through the best choice of means (and which is especially suited to market and bureaucratic social orders).

Three kinds of orientation to the external societal context are possible for alternative forms of education: *separation*, *engagement*, and *activist*. These are set out here as useful tools for analysis rather than being proposed as a comprehensive conceptual framework. The point is not to classify the alternatives in this volume, as the orientations are not mutually exclusive nor necessarily competing stances. Rather, they are put forward as a helpful analytical device for highlighting differences of emphasis amongst alternative types of education.

One type of orientation is *separation*. This involves the alternative type of education maintaining a distance and its own protected environment. It does not necessarily preclude trying to have some influence on other educators and educational systems or being open to learning from other forms of education, but the main emphasis is on creating and sustaining a separate educational environment. External boundaries are therefore strong. Examples where there is a notable separateness in this sense include Steiner/Waldorf schools in contexts such as the UK (chapter 11); Summerhill, which displays powerful, invisible, boundaries marking out school pupils and "downtowners" (local inhabitants) (chapter 3); and Islamic schools, which are seen by many Muslim parents as providing educational settings that protect their children from secular influences in the public schools and society, and cultivate a strong religious identity (chapter 6).

The second orientation to the societal context is that of *engagement*. This is aimed at mitigating adverse consequences of separation and/or influencing the wider social and educational environment by building pragmatic relationships with, for example, mainstream education and governments. A strategy of engagement may lead to some alternatives being *accommodationist*, that is, practising an alternative education form but at the cost of retaining or even reinforcing key elements of the dominant, instrumentally driven culture. Many Islamic schools in the United States, for example, are drawn to give attention to "secular" goals (such as attainment of academic qualifications) because of parents' ambitions for the success of their children after leaving school. The entry of a Steiner/Waldorf school in England entering the publicly funded state sector raises issues concerning whether this may erode some of its fundamental principles, such as the avoidance of tests and assessments that are integral to the UK Government's educational policy in England. Inherent to engagement is the need for compromise.

The third type of orientation is an *activist* one. This involves trying to alter the conditions that give rise to any adverse consequences suffered by the educational alternative from its positioning on the margins, and/or to change the policies and practices that restrict mainstream education. The aim is a wider social change, in which education is an inherent part. This is apparent with Māori education (chapter 8), and with developments in First Nations education in Canada, which explore what happens when Indigenous ways of knowing and being in the world are brought to bear on Eurocentric forms of education and schooling (chapter 9). The educational reforms in Latin America (chapter 2) are aimed at system-wide change, creating participatory educational arrangements very different from commodified relationships in the economy. The K20 Center in its model of systemic change in Oklahoma (chapter 1) also aims to contribute to the transformation of inequitable conditions through the creation of democratic learning communities. As noted, the case of Summerhill is more of an isolated (separation) orientation and not part of a formal movement, though it has acted and continues to act as a beacon and example of progressive change that influences other educators.

Arguably, as distinct, separate sites of education, many of the alternatives in this volume, including Summerhill, Quaker Schools (chapter 4) and the Dharma School (chapter 5), contribute to the wider society's capacity for democracy. Their ability to do this rests on the extent to which they are successful in educating in ways that develop students' sensitivities, identity and self-esteem, as well as their cognitive abilities and skills, so that they leave school capable of being independent, free-thinking individuals—the explicit goal of, for example, Steiner/Waldorf education. Raymond William's distinction between alternative and oppositional, therefore, may be a fluid rather than a rigid boundary. However, a great deal more research is needed on the actual impact of different kinds of alternative education on student development before conclusive assessments can be made on the extent to which they are successful in nurturing rounded student development and contributing to society's capacity for democracy. And such research would need to take account of the differing social positionings of alternatives that results in their tending not to serve a cross-section of students, according to such variables as social class, ethnic identity, and so on, but to draw from a particular group—middle-class families in some cases, such as Summerhill and Quaker schools, or socially and economically disadvantaged communities in others, as with Māori and First Nations education.

Waitere and Court's challenging of the idea of understanding Māori education as an alternative, at least in any simplistic sense, is particularly interesting. They construct, as the third element of their three-part typology of alternativeness, a concept of the (alter)native, which describes the direction of travel of Māori education, as education that is both distinct from and in an equal, interactive relationship with the education system created by the colonizing inhabitants of Aotearoa New Zealand. It is a reminder that educational alternatives are formed and situated in a context where resources and opportunities are unequally distributed. Bekerman highlights the effects of the power relationships in the wider society when he observes of the Palestinian Jewish schools (chapter 7) that "Context might be too powerful to be overcome even by the most well-intentioned bilingual educators" onto whose shoulders fall the pressures to put into daily operation an alternative that cannot in itself solve the wider contextual problems. In the Latin American reforms aimed at promoting democratic involvement of nonprofessional stakeholders, Gvirtz and Minvielle's research found that formal participation tended to be "peripheral" or "passive" because of basic assumptions "deeply settled in the school culture," which meant that automatic deference was given to professional educators.

> **Reflective Activity Box (I)**
>
> The different modes of orientation can be used by educators to reflect on their positioning and those of their institutions and wider communities and context.
>
> For alternative educators, one question for reflection is the extent to which they (or their institution or collective group of schools) practices *separation, engagement,* or an *activist* stance toward mainstream education and other alternatives. How could this stance be improved, and who would benefit from this?
>
> An issue for mainstream educators to consider is how they (or their institution, community of schools, or local/national administrative authority) view and relate to alternative types of education: Do they see the alternatives as practicing *separation, engagement,* or *activist* modes of orientation? Does their own mainstream context keep itself separate from alternatives or does it seek to engage with them? How could relationships be improved, and who would benefit from this?

Diversity for Democracy

What role can alternatives play in a democratic, multicultural society? The idea of diversity amongst schools challenges the notion of the common school. The goals of the common school, summarized by Halstead (2007), are to develop commitment to a set of shared values and a unified sense of citizenship; reduce the effects of socioeconomic disadvantage and provide equal opportunities; offer a common educational experience to all children; and prepare children for life in a multicultural society by providing experience of living and working alongside children from cultures other than their own. This is not the place to rehearse and review the arguments for and against a common school system (see, for example, Haydon, 2007, Pring, 2007, Rofes and Stulberg, 2004). However, we want to foreground some of the arguments for diversity and difference amongst schools. For example, first, where the cultural identity of a certain group in society is of compelling importance to that group, there is a case for its maintenance in protected forums of education on democratic grounds—because of its significance for that group and because loss of identity and meaning impoverishes people's lives (Berger, 1977).

Second, following a different, complementary line of argument, Alexander (2007, 621) advocates what he calls the "pedagogy of difference," which is based on recognizing the value of education within a particular culture. If children are all required to be educated into "a comprehensive liberal view of the world" (620), this, argues Alexander (2007, 620), is another form of imposition by the state. Instead, he draws attention to the benefits of specific cultural forms of schooling that educate children into an identity and which aim to nurture them not simply into becoming "rationally autonomous beings" but as "independent moral agents, free within limits to choose between several life paths and capable of distinguishing between them" (Alexander, 2007, 621). Moreover, the pedagogy of difference requires

> not only a deep immersion in the stories and practices of the tradition into which one is being initiated, but also opportunities to learn of other traditions and to experience them as well... (Alexander 2007, 621)

As Alexander (2007, 621) summarizes it:

> To understand myself I must encounter the other; but to genuinely encounter the other I must also understand myself.

Examples of sustained education into identity include Islamic schools, which many Muslim parents choose for their children, amongst other reasons, in order to protect their children from secular influences and cultivate a strong religious identity. Other alternatives in this volume also can be seen as fostering identity—strong cultural identities, as in the case of Māori and First Nations education. In the case of Steiner/Waldorf education, it is not a matter of nurturing a religious or cultural identity; rather, the aim is to develop a secure, matured independent self-identity. In the Palestinian Jewish schools, the identity fostered is one that values both cooperation and mutual understanding between cultures and strong identification by each student and their family with their own culture. However, Bekerman observes that something of the depth of the individual cultures is lost: "shadows of [the cultural traditions] in the shape of truncated holy texts or cuisine recipes make their appearance in the school scene," reducing them to a "reified folkloristic perspective." This is a cautionary observation that cultural messages are mediated and affected by the character of the educational contexts in which they are sought to be conveyed.

Third, diversity facilitates innovation and experimentation. This argument is based on the belief that there are multiple pathways to fostering freedom and the capacities to exercise it, and that it is right for these to be given expression in different kinds of schooling. There are diverse ways of developing the spiritual and aesthetic sensitivities, the secure identity and self-esteem, and the cognitive abilities and skills necessary to nurture independent-minded individuals with a sense of civic and ecological engagement. Various alternatives have their virtues in this regard—from Quaker schools and the Buddhist Dharma School, to secular Summerhill; and, whilst none is perfect, there are things that can be learnt from each. The potential for mutual learning across schools about matters concerning the deep purpose of education and the fostering of democratic citizenship is increased where alternative forms of education are facilitated and valued. Opportunities for innovative types of schools can lead to the creation of educational sites that form "mechanisms of resistance" and that interrupt the overwhelming drive of much of the institutionalized education system "toward consistently reproducing the status quo" and social inequalities (Rofes, 2004, 256). Nevertheless, such diversity needs to be embedded in a policy framework and education system that seeks to apply protective principles, as discussed in the Introduction, advance social justice, and achieve key aims associated with a common school system—namely, fostering commitment to a set of shared values and sense of citizenship, reducing the effects of socioeconomic disadvantage, and preparing children for life in a multicultural society.

In our view, diversity and difference amongst schools has, therefore, enormous potential for interinstitutional learning. It gives scope not just for educating in traditional identities that are important for some people and groups, but for innovation and experimentation in progressive education. This is how innovative models of schooling such as Summerhill, Steiner/Waldorf education, and the Palestinian Jewish schools have come about.

If such diversity is to achieve its promise for learning and development across education, and not to be a confined local change, inward-looking, or socially divisive, it needs to be part of an education system that practices *intrinsic co-operation* (Woods and Woods, 2002), that is, cooperation between diverse (alternative and mainstream) schools in which different conceptions of valued learning are given expression, respected, and shared. In a system that practices intrinsic cooperation, different conceptions of valued learning are valorized (intrinsically for themselves; not as techniques to better measurable performance), and there is a commitment to understanding the aims, ideals, and practices embedded in other educational settings. Intrinsic cooperation values social bridging both within schools—where it enhances improvement through "teachers working democratically, horizontally, to share practice, observe and learn from one another" (MacBeath, 2004, 45)—and across schools. More specifically, intrinsic cooperation involves commitment to interactional processes that comprise (Woods and Woods, 2002):

- *dissemination* (creatively applying in one educational setting ideas and practices of another type of school or educational philosophy),
- *critical reflection* (willingness of educators in one school type or educational philosophy to reflect critically on their own ideas and values in light of their awareness of other school types and philosophies),

- *integration* (shared, positive identification as a common educational community on the part of educators working in different types of school or educational philosophy).

Integral to these processes is cultural networking—that is, understanding symbols, meanings, and customs transmitted within and across communities (O'Hair and Veugelers, 2005, 3). What is key here is that boundaries between alternatives and mainstream, and between alternatives, are permeable. Indeed, the alternatives, as in Waitere and Court's conceptualization of the "(alter)native," cease to be separate. There is fluidity and recognition of interconnectedness. And as the insight into First Nations education reminds us, recognizing "the relations of all beings to one another" and "placing the self with one's relations for the benefit of all of creation" are fundamental starting points for understanding education.

> **Reflective Activity Box (II)**
>
> For the type of cooperation discussed here to grow, educators in all types of education need to work on ways of expanding opportunities for mutual learning across educational contexts—in other words, to increase the chances for people to experience "connectedness" between different settings. Reflecting on basic orientations, as suggested in Reflective Activity Box (I) at the close of the previous section, is one way to begin. This could be followed up by educators reflecting on the extent to which they (or their institution, collective group or local community of schools, or local/national administrative authority) practice the interactional processes outlined above:
>
> - *dissemination* (creative application of ideas from other settings),
> - *critical reflection* (willingness to be self-critical in the light of other school types and philosophies),
> - *integration* (shared identity with other educators across different types of educational context).
>
> Which of these describe your, or your context's, current practice, and in what ways are they put into operation? In what ways could current practice of interactional processes be improved, and who would benefit?

Moving Toward Developmentally Democratic Sites of Education: A Framework

Much of contemporary education has a great capacity to keep "running on the spot,"[1] or, we could say, running faster and faster just to stand still in the increasingly competitive global market place. This is the fate of the newer and growing type of educational organization driven by bureau-enterprise culture (Woods, 2007a), which seeks to create a dynamism and capacity to innovate continually in similar ways to the private sector but does not inherently define itself in terms of the deep purposes of education. Bureau-enterprise culture is an organizational order characterized by a dominating instrumental rationality orientated to the market-driven economy, which is embodied in flexibilised sets of organizational roles, a more entrepreneurial and consumer responsive culture, a belief in flatter hierarchies, a visionary zeal to achieve system and organizational goals, and systematic monitoring and measurement of performance in formally rational ways. Bureau-enterprise culture runs faster and continually innovates to achieve and improve on targets and to create the type of educational experience viewed as necessary for today's competitive society. It is part of what Ball (2007, 136) refers to as the "re-narration of schooling and learning through innovation and enterprise," which is "part of an adaptation to a particular version of the economic and social context of globalization and post-modern society."

Yet many educators in mainstream education abhor the performative culture and work to forge a more progressive education environment. They want to be "on the road to somewhere," to echo the title of chapter 7 on Palestinian Jewish schools.

In this section we set out a framework for reflecting on the direction and foundations for such a road—a road toward developmentally democratic sites of education. The features of developmentally democratic educational settings are not an amalgam of the characteristics of the alternatives discussed in this volume. They are, rather, our suggestions of what can be drawn from a consideration of the alternatives, and thus arise from a particular perspective. The framework that we have constructed, shown in figure 12.1, comprises overlapping features of the kind of education that strives for innovative ways of creating an ethos, curriculum, and pedagogies that are inclusive, open to alternatives, educate for and within democratic society, and nurture rounded physical, spiritual, emotional, cognitive, ethical, and social development in students. In practice, the relative weight, pattern, and nature of these features will differ between particular educational settings in the light of their histories, contexts, and stakeholders. The framework of features is intended to be relevant for reflection on both existing mainstream and alternative educational institutions, by practitioners and policy makers seeking areas for improvement, and on plans and possibilities for new educational institutions.

Reviewing the alternatives has helped in further delineating what the notion of developmental democracy (Woods, 2005) could look like in educational practice.

The ideal-type bureau-enterprise offers a way of throwing into sharp relief the elements of developmentally democratic educational settings.

Bureau-enterprise	→	Developmentally democratic
Performative framework	→	Holistic firm framing
Goal-driven technical approach	→	Integrative approach for freedom
Instrumental practice	→	Mindful practice
Technical co-operation	→	Intrinsic co-operation

Figure 12.1 Features of Bureau-enterprise and Developmentally Democratic Education

Hence, in figure 12.1 these are contrasted with features of the type of education characterized by bureau-enterprise culture. Both types—the developmentally democratic and the bureau-enterprise—are represented as ideal-types in the figure. Developmentally democratic features, as given in the discussion below, are illustrated through examples drawn from the alternatives in this volume.

Holistic Firm Framing

One of the organizational features reinforced by the alternatives is the importance of some kind of strong boundary that frames the school, some type of firm framing (Woods, 2005). However there cannot, of course, be any kind of firm framing. The requirement is for what we term here holistic firm framing—or a transformative alternative framework.[2] This is a vision, a set of principles or philosophical underpinning and overarching institutional arrangement that explicitly signals the multidimensional nature of people and affirms the commitment of the school to social justice and to respecting this in its curriculum and pedagogy and to developing in its students the capacities for freedom, relationships, and "perceptual/intuitive awareness" (highlighted in the discussion of Montessori education in chapter 10) necessary for active democracy. In particular, such firm framing expresses the necessity for balance that gives to the spiritual its appropriate prime position and places at the centre a concern with ultimate reality in some sense. This is integral to the rounded development of students, which embraces the physical, spiritual, emotional, cognitive, ethical, and social dimensions.

Holistic firm framing contrasts with the performative framework of bureau-enterprise culture, which focuses principally on the metrics of attainment and the analysis of improvement (or otherwise) in terms of measurable change. Performativity ties education to "the information systems of the market and customer choice-making and/or to the target and benchmark requirements of the state" (Ball, 2007, 27).

The orientation given by holistic firm framing, which defines the ethos of the school and points to the identity being nurtured and encouraged in the members of its community, is different. It involves, at least in part

> recentring the culture of schools so as to encompass a shared vision and values orientated towards democratic ideals and practice. Moreover, the point is not for it to be just a vision that is set down in text, but one that is constituted through everyday dialogue and, hence, is part of the "creative fashioning" of educational discourses. (Woods, 2006, 331)

The K20 Model provides a distinctive example of a substantial framework for schools, which includes a vision focused on creating democratic learning communities, university-based institutional support through the K20 Center, connecting with over 500 schools, and sets of principles and tools for practice, such as the *IDEALS* framework (representing Inquiry, Discourse, Equity, Authenticity, Leadership, and Service).

A strong holistic framework is one dimension of one of the dualities of democracy—that of embracing the ideas of firm framing and free space (Woods, 2005). This duality encapsulates the idea that democracy involves both a definite structure that provides, on one hand, a sense of position and place in an organization, concepts, ideas, and a context of values to relate to, and a rhythm of social relationships; and, on the other, loose-structured creative social areas where hierarchy and assumptions of knowledge, norms, and practice are minimized—with movement between these. It can be seen as the operationalisation of the substantive and protective principles. So, for example, in Summerhill there is the Meeting that makes laws collectively binding upon everyone, whilst individually students are able to decide if they wish to have time free from lessons (that is, they can create their own free space if they perceive they need or want it) and negotiate the weak organizational boundaries of the school. Alternatives such as Quaker schools, the Dharma School, and Steiner/Waldorf schools all have elaborated philosophical/spiritual principles and worldviews—which is not the same as requiring adherence to fixed dogma. In their different ways, these alternatives combine detailed propositions about values, meaning, and human development, whilst also inviting interpretation and criticism.

Integrative Approach for Freedom

An integrative approach for freedom encompasses both a method and orientation (*integration*) and a guiding aim (*freedom*). The importance of integrative approaches is a key theme of the organizational, curricular, and pedagogical features of many of the alternatives in this volume. The fundamental purpose of these integrative approaches is the development of freedom. The understanding of freedom in this context is not defined by or confined to consumer choices amongst commodities, but, rather, is about the capabilities that enable the person to choose and behave as a moral agent and develop and act upon their multidimensional potentialities.

This contrasts with the technical approach of bureau-enterprise culture, which involves the instrumental use of both fragmented and integrated approaches and of innovative ways of working as they appear to suit utilitarian requirements and specified goals.

Integrative approaches can be manifest in a number of ways. First, there is *curriculum integration*, in which the analytically distinct dimensions of personhood are integral to the educational experience offered, and none is marginalized. This is to the fore in most of the alternatives and is a theme that is variously expressed in practice. It is particularly exemplified in Quaker education (where the academic side of life is respected but it is the development of the whole personality that is the main goal, including the encouragement of artistic and practical skills), the Dharma School (in which there is an emphasis on progressing autonomy, enquiry, and questioning, as well as virtue, compassion, and spirituality, affirming the holistic nature of personhood and our relationship with the world), First Nations education (where the relationship of all things is fundamental to the world views of the peoples), Montessori education (which is about educating the senses as well as the mind), and Steiner/Waldorf schools (which encourages balanced growth toward physical, behavioral, emotional, cognitive, social, and spiritual maturation and the curriculum remains broad right up until age 19). The perspective underpinning First Nations education gives special emphasis to relations through the image of the circle. Haig-Brown and Hodson articulate this through two patternings:

- the medicine wheel showing the four aspects of the self-in-relation (physical, emotional, intellectual, and spiritual),
- the six interrelated directions of education: spirit, land, regeneration, remembering, resistance, responsibility.

The second way of manifesting an integrative approach is through *pedagogical integration*. This is encapsulated in the concept of *embodied education*, defined by Rathunde in his discussion of Montessori education, as education in tune with the intimate connection of the body and the mind. It is exemplified in "flow," which is action that involves deep engagement and interest, characterized by deep concentration, merging of action and awareness, changes in the awareness of time, feelings of clarity and control, a lack of self-consciousness, and feelings of intrinsic motivation. A pedagogical implication of seeking to facilitate embodied education is the importance of creating what Rathunde calls the *prepared environment*, which involves three aspects: finding the appropriate balance of skills and challenges, freedom and discipline, and body and mind.

The concept of embodied education works with another of the dualities of democracy—rational capacities and affective capacities—which combines development of calculative and cognitive capabilities with fostering of noncognitive dimensions (Woods, 2005). The notion of embodied education offers a valuable orientation to pedagogies; it also raises the question of what are the faculties and dimensions that make up the body, to which the mind needs to be attuned. Embodied education clearly includes the physical aspects of the body, the five senses, and the affective dimension that incorporates feelings. What we want to make as an explicit addition to these is the faculty for spiritual development. If disembodied learning is seen as Type 1, and Type II involves embodied learning that recognizes the importance of physical and emotional experience as well as cognitive faculties, there is, we argue, a *Type III embodied learning* that extends the understanding of the body to embrace the intuitive and spiritual capabilities inbuilt to the person, which nurture and enhance an inner state that is harmonious and in tune with a transcendent reality and the *supreme value* of connectedness and compassion (Woods, P. A., and Woods, G. J., 2008).

A third way of manifesting an integrative approach is through what we refer to here as *emplaced education*—education that recognizes and values the close relationship between the embodied learner and the environment. In First Nations education, the beginning orientation is to the land. Always reinforcing the relations of all beings to one another, Haig-Brown and Hodson explain that the land becomes the first teacher, the primary relationship, and is to be understood as a complex being—a spiritual and material place from which all life springs. The connection to the land is important for the Māori peoples too (tangata whenua: people of the land).

The notion of emplaced education embraces the importance of connection with the land and beyond that to the space and relationships people are grounded in. This is illustrated by the significance of out-of-classroom learning that Stronach and Piper draw attention to at Summerhill. They highlight the social learning that takes place through informal learning sets, based on a myriad of relationships—teacher-pupil, pupil-peers, mixed-age groupings, and so on. Strength and quality of the relationships experienced by students is an important feature of emplaced education. Ashley highlights the stability offered by Steiner/Waldorf schools' provision for a sustained relationship between class teacher and class over several years. Relationships are clearly a strong feature too in Summerhill, though in a different way. There, continuity of relationship is not focused on a class teacher but is constituted by networks of relations within the community.

The idea of emplaced education can be seen as enfolded within the concept of *relational consciousness* (Hay with Nye, 1998), which is characterized by a distinctively reflective consciousness, or meta-cognition, entailing some degree of awareness of one's own mental activity and a sense of relationship with others, the self, the world, and (for some) God (Hay with Nye, 1998, 113–114). It is a heightened sense perception that reduces for some period the "psychological distance" between the person and the rest of reality (Hay with Nye, 1998, 18). Emplaced education draws attention to the situatedness of the embodied learner—the relationship with others and the world, and the transcendent, which are part of relational consciousness.

The positioning of the relationship with the transcendent crosses over both embodied and emplaced education. Transcendent power, theorized by Glenys Woods, following work by Sir Alister Hardy and others (Hardy, 1966, 1991; Hay 1987; James, 1985; Woods, G. J., 2007), is experienced as part of and beyond the person. It is sensed by the person and characterizable as a particular kind of inner strength (so a feature of Type III embodied learning), and experienced as a beneficent power far greater than the individual self (so a feature of emplaced education). Evidence suggests that this type of quintessential spiritual experience is widespread amongst the general population, including educational leaders (Hay, 1987, Hay and Hunt, 2000, Woods, G. J., 2003, 2007). If this is to be appropriately acknowledged, the spiritual requires central recognition in the integrative approach. Such a central placement is clearly understood within First Nations education, for example. In Haig-Brown and Hodson's chapter, an example is given of an experience of expanding consciousness that gradually takes the form of a felt relationship with all creation—which, moreover, is not only an individual experience of personal significance but also a repetitive theme among Aboriginal peoples. Openness to such expanding awareness is also a vital part of alternatives such as Quaker education and the Dharma School. Additionally, contemporary research on the incidence of quintessential spiritual experience emphasizes that such awareness is not confined to those with religious beliefs but is consistent with "secular" worldviews, such as humanism, agnostic perspectives, and atheism (Woods, G. J., 2007). The conclusion we draw is that recognition of *spiritual connectedness* is a vital component of the integrative approaches encapsulated by the concepts of embodied and emplaced education, and that it has relevance and acceptability across a wide range of alternative and mainstream schooling.

A fourth way of manifesting an integrative approach is through the operationalisation of *democratic rationalities*, which include ethical rationality (shared pursuit of truth and meaning), discursive rationality (exchange and exploration of views and open debate), and decisional rationality (active contribution to the creation of the institutions, culture, and relationships people inhabit) (Woods, 2005). These are facilitated through weak boundaries operating within a strong framework—holistic firm framing. Summerhill provides a strong illustration of the operation and wide involvement of staff and students in discursive and decisional rationalities. The aim of opening out discursive and decisional rationalities is evident in the Latin American reforms and the work of the K20 Center in Oklahoma. The K20 Model aims to integrate the educational imperative of the school, the engagement of students (by enhancing the relevance of learning), and sustained professional development through a vision of democratic learning communities. One of the ways this is manifested is through working toward democratic classrooms where students

> have frequent opportunities for input and decision making about school and classroom life...share ownership and responsibility for classroom and school property, space and environment... [and, for example] develop and evaluate classroom rules, procedures, routines, and activities and are involved in addressing social problems that occur during the schoolday. (O'Hair et al., 2000, 339)

Implicit in this are deeper issues of values and meaning that are raised by practical questions and challenges, which take participants toward the ethical rationality that involves the individual and shared search for better understandings of the world (scientific, social, psychological knowledge, for example) and of questions of purpose and value. This ethical rationality in our view is essential to the nature of the truly democratic learning community, which is characterized by "change and innovation [that] is measured not by narrow performance criteria but by an understanding of what it is to be a developing human being and the kind of organizational environment that nurtures this" and by "the development of the inner life of the critical-democratic citizen, as much as a dispersal of decisional rationality in which power and authority are shared" (Woods, 2007b, 40).

Not all the alternatives espouse democratic aspirations in the way they run themselves. A key question for educators is how to draw the democratic geography of their schools. Steiner/Waldorf schools are run by a collegiate of teachers (thus, in principle, engaging each teacher in the democratic rationalities), but this does not extend to students and there can be tensions with regard to membership of nonprofessional staff in the collegiate system

(Woods and Woods, 2006a/b; Woods, G. J., and Woods, P. A., 2008). And, as is evident from the examples of the Latin American reforms and the experience of Steiner/Waldorf schools, for example, the practice and implementation of democratic participation is difficult and often far from perfect. The experience of Minas Gerais, which Gvirtz and Minvielle highlight, shows that some opportunities to enhance participation can be found even if the ambitious formal aims prove difficult to achieve. In Minas Gerais, more success for nonprofessional participation was found through less formal participation spaces. And it is possible to see how space for discursive and ethical rationalities can be created in schools that do not style themselves democratic in their organizational structure. So, for example, Merry and Driesson draw attention to the findings of Istanbouli (2000) and Oriaro (2006), which show that debate about Islam, questioning of authority, and exposure to alternative views are possible and apparent amongst students in some Islamic schools.

Ashley, writing on Steiner/Waldorf education, reminds us that the aims of mainstream education are often couched in terms of multiple dimensions of the person, articulating, as in England, the importance of social, moral, spiritual, and cultural development; but the question is how this aspiration is translated into practice. The integrative approach to curriculum, pedagogy, emplaced education, and democratic rationalities is intimately connected with the goal of freedom, since the capability to exercise freedom is nurtured by developing a balance between the dimensions of the person. As Haig-Brown and Hodson observe, students may come to know a body of transmitted knowledge or a set of skills, but this does not mean that they know how to learn or how to live in freedom.

Mindful Practice

Mindful practice is closely related to the integrative approach. It describes the ongoing, dialectical interrelationship of inner practices and outer activities, practiced and developed by both staff and students. Democratic learning communities outwardly aim to enact the ideals represented by the democratic rationalities and "inwardly—within people—... [give] priority... to the full development of human potentialities and the importance of meaning and purpose" (Woods, 2007b, 41). As Henderson and Slattery (2006, 4) put it, struggle with oppression, which is integral to the building of a democratic society, "is linked to the struggle with repression": in the "dialectic of inner and outer change," specific attention has to be given to the inner dimension of democratic freedom that "is the embrace of a diversified psychic movement through an open-hearted attunement to the eros—the Tao, or loving way—of democratic pluralism" (Henderson and Slattery, 2006, 3).

Spiritual connectedness, discussed above in relation to embodied learning and emplaced education, is integral to mindful practice. There is evidence of spiritual reflection being a significant aspect of meaning-making amongst staff in a diverse range of modern organizations (Casey, 2002). However, where the ideal-typical bureau-enterprise culture differs from the conception of mindful practice is in the dominance given to the instrumentality of practice. Attention to inner development, in the sense of right attitudes, emotional intelligence, and even spiritual development, can be recognized and encouraged in bureau-enterprise culture, but this is focused on the impact it is perceived to have on performance of self and of the organization. In the ideal-typical construction of bureau-enterprise culture, there is an absence of intrinsic valuation of inner development and outward activity. Programmes "currently extolled by organization culturalists and management motivators now overtly encompass the *utilization* of religio-affective impulses and noneconomic and nonscientific rationalities…" (Casey, 2002, 160; emphasis added).

By contrast, mindful practice is intrinsically valued. It can be put into operation in different ways: in Quaker practice, as attentive silence, which waits for and allows inspiration to come, and teaching, which makes way "for the softening influence of divine goodwill in [students'] hearts"[3]; in Buddhist practice, as silent observation, which is practiced to create positive qualities, and the development of mindfulness, which aims to nurture self-awareness and acceptance of others as they are and in habits of being positive in their thoughts toward them; in Islamic schools, as mindfulness of God; and as the widespread phenomenon of transcendent awareness that is integral to quintessential spiritual experience.

What these practices point to are attempts to give a depth to the development of self-awareness. Self-knowledge comes in many ways. At Summerhill, Stronach and Piper highlight how students are good at reading each other and intuiting boundaries, as well as being experts on "themselves." In his discussion of Montessori education, Rathunde explains the relevance of the concept of *psychological complexity* (Rathunde and Csikszentmihalyi, 2006), which refers to the self-regulative capacity in a person to find flow. This capacity is more likely to be developed in the young person if they have spent time in the right context—the prepared environment, which provides proper balances of skills and challenges, freedom and discipline, body and mind, and so on—so that they are able to recreate moments of deep engagement on their own and stay on a path of lifelong learning, contributing to their holistic development.

However, if relational consciousness is to be as rounded and full as possible, the inner dimension and the connection through that with the spiritual has to be a part of educational practice.

We describe mindful practice as dialectical, pointing to the fact that the inner activities or inner work are not navel-gazing but interact with outward activities: as it does, for example, with Steiner/Waldorf teachers in their pedagogy and their participation in decisional rationality in the collegiate of teachers (Woods, G. J., and Woods, P. A., 2008), and as with the Quaker approach in which contemplative silence complements commitment to social action and justice. The inner life is also connected to outer relationships through shared ceremonies and events. Haig-Brown and Hodson (chapter 9) observe that it is partly through ceremonies that we can connect to the spiritual reality of an educational journey.

If participative democratic rationalities are working well, we would suggest that this is likely to have a beneficial impact on feelings of well-being and confidence amongst those who are involved. To put it another way, developmental democracy involves a therapeutic rationality whereby people are empowered and enabled by the participative culture and structures of the organization (Woods, 2005). Integral to this is mindful practice that comprises both the inner work of the person in developing awareness and outward symbolic expressions of what unites and is considered important to the educational community.

Rituals and routines are particularly significant in Steiner/Waldorf schools, for example, both for students and staff. The following is an example of a morning verse as recited in a Steiner/Waldorf school, in classes 5 to 8 (Woods, P. A. et al., 2005, 205). The significance of these words is not from any ideology or philosophy that they represent, but the shared orientation they offer.

> *I look into the world*
> *Wherein there shines the sun*
> *Wherein there gleam the stars*
> *Wherein there lie the stones.*
> *The plants they live and grow*
> *The beasts they feel and live*
> *And man to spirit gives*
> *A dwelling in his soul*
> *That living dwells in me*
> *God's spirit lives and weaves*
> *In light of sun and soul*
> *In heights of world without*
> *In depths of soul within*
> *To thee, oh spirit of God*
> *I seeking turn myself*
> *That strength and grace and skill*
> *For learning and for work*
> *In me may live and grow*

Similarly, the following—one of the verses often used in opening teachers' meetings in Steiner/Waldorf schools (Woods and Woods, 2006b, 9)—encourages individual and communal reflection toward a shared point of reference.

> *Imbue yourself with the power of imagination*
> *Have courage for the truth*
> *Sharpen your responsibility of soul*

Intrinsic Cooperation

Intrinsic cooperation values external collaboration with other educational institutions and opportunities to learn from differences, not only in relation to pedagogical techniques and the like, but also through engagement with differing perspectives on the deeper purpose and philosophies of education. It is contrasted with technical external cooperation, where the bureau-enterprise type of school cooperates principally for the benefits it brings in increasing standardized measurable criteria (Woods and Woods, 2002). The exclusive aim with technical cooperation is to extract "collaborative advantage" (Huxham, 2003).

Intrinsic cooperation is embedded in a sense of identity and common purpose with other educators, expressed through interactive processes such as critical reflection and integration. These concepts and the importance of intrinsic cooperation where there are diverse types of school were discussed above. What emerges from thinking deeply about the variety, nature, and experiences of alternative types of education is the significance of boundaries—where they are drawn strongly (around the principles and key pedagogies of an educational approach, for example), and where they are more permeable and weak (enabling participation and critique, for example, and foregoing dogma). For mutual learning and reflection to take place, neither alternatives nor mainstream can be in the *separation* mode of orientation. The need is to recognize and reinforce the principles that alternatives and mainstream share and that transcend boundaries, such as a commitment to social justice, to education for the full development of the person, and to the overcoming of historical and contemporary inequalities that restrict the educational opportunities and experiences of certain groups. Alternatives from this perspective are not options alongside mainstream, but, rather, *activist* educational approaches interacting with and raising questions of other approaches and unjust power structures. The development of Māori, First Nations, and Palestinian Jewish education are insightful examples in this regard, though other alternatives also probe and question and contribute to change. Equally, as we have emphasized in the Introduction, learning and engagement is a two-way process between alternative and mainstream.

The logic is away from a fundamental boundary between alternatives and what for convenience we call mainstream education. The common focus must be how to provide true education for the growing person in societies in which powerful socioeconomic forces tend to instrumentalise the individual and dissect their attributes, skills, and capabilities. What is important is not so much creating the *common school* but ensuring that there is a unifying *common purpose* embracing education across schools.

Reflective Activity Box (III)

The framework of developmentally democratic features, summarized in figure 12.1, is designed as an aid to reflective practice, which can be used for reflection at different levels: individual; institution; collective group or local community of schools; local administrative area; national administrative area. The features can be used as overarching themes orientating reflection, and/or more specific concepts may also be used. All educators might use the framework as an audit, to consider the extent to which their educational setting achieves the following (selected concepts and issues are listed for each of the features):

- holistic firm framing
  - balancing of beliefs/principles about values, meaning, and human development, with scope for interpretation and criticism
- integrative approach for freedom
  - curriculum integration (physical, spiritual, emotional, cognitive, ethical, and social dimensions)
  - pedagogical integration (embodied education)
  - emplaced education
  - democratic rationalities (ethical, discursive, decisional)
- mindful practice
  - spiritual connectedness
  - psychological complexity and "flow"
- intrinsic cooperation
  - outward orientation: separation, engagement, activist (See Reflective Activity Box I.)
  - interactional processes: dissemination, critical reflection, integration (See Reflective Activity Box II.)

Connectedness

A preeminent theme in this volume is connectedness. What the best of alternatives and progressive education in the mainstream point toward is something in the human spirit that arises from deep within—an awareness that too much education fails to aspire to create the best conditions in which the human being can connect with their highest potential. They provide lessons in imperfect struggles—that is, attempts, with greater or lesser success, to fashion the best environment for holistic growth. The Sufi poet, Rumi, directs attention to the importance of detaching from those things that obstruct unity with that which is of supreme importance and value.

> *Take away what I want*
> *Take away what I do*
> *Take away everything*
> *that takes me from you*[4]

To put this in a more elaborate way:

> To release us from the prison of fear, we must seek out that which creates and sustains the desire to live without dependence, or submission to any authority. There can be no limited perspective or particular exemptions. Liberation must be total, with nothing partial in its resolution. As such it cannot be simply a passive reconciliation. It must be active, dynamic, always seeking interaction with another to make itself complete, and consequently expanding its own potential. (Keenan, 1997, 16)

Education is, ultimately, about dissolving boundaries—between institutional religion and secular systems; between alternatives and mainstream; between people, where injustice and conflict distort relationships. If we work with the poetic extract at the start of the Introduction, the black bird it mentions is "impossibly distant" because, as we interpret it, it is not far away in the sky but is that creative life force, love, that is within and between the people running on the ground. What is "possibly near" is the creation of opportunities for stillness and calm, in ways appropriate to context and circumstances, to realize relational awareness, to detach from all that stands in the way of real progress, and to provide the grounds for action and change that respects, involves, and nurtures people in their full humanity and contributes to a more democratic and socially just world.

Less running on the spot. More pathways to enlightened learning.

Notes

1. Like the small creatures in the poem extract at the beginning of the Introduction.
2. In chapter 9, Makere Stewart-Harawira is quoted as expressing the need for "transformative alternative frameworks," in relation to traditional indigenous social, political, and cosmological ontologies.

3. Dr. Fothergill, founder of the first Quaker boarding schools, quoted in Walvin, (1997, 95).
4. Source: "Spirituality as Service," by Pierre Pradervand, *Cygnus Review*, May 2008, 10.

Bibliography

Abram, D. (1996). *The Spell of the Sensuous: Perception and Language in a More-than-Human World.* New York: Vintage Books.
Alexander, H. (2007). "What is Common About Common Schooling? Rational Autonomy and Moral Agency in Liberal Democratic Education." *Journal of Philosophy of Education,* 41/4: 609–624.
Alexander, H. A. (2003). "Moral Education and Liberal Democracy: Spirituality, Community and Character in an Open Society." *Educational Theory,* 53/4: 367–387.
Alfred, T. (1999). *Peace, Power and Righteousness: An Indian Manifesto.* Toronto, ON: Oxford University Press.
Al-Haj, M. (1995). *Education, Empowerment, and Control: The Case of the Arabs in Israel.* Albany, NY: SUNY Press.
Amara, M. H. (2005). *Summary Report: The Bilingual Model. Jerusalem: Hand in Hand-The Center for Arab Jewish Education in Israel.* Internal unpublished report.
American Friends Service Committee (1955). *Speak Truth to Power: A Quaker Search for an Alternative to Violence.* www.quaker.org/sttp.html. Accessed March 25, 2008.
Ames, C. (1992). "Classrooms: Goals, Structures, and Student Motivation." *Journal of Educational Psychology* 84: 261–271.
Anderman, E., Maehr, M. and Midgley, C. (1999). "Declining Motivation after the Transition to Middle School: Schools Can Make a Difference." *Journal of Research and Development in Education,* 32/3: 131–147.
Anderson, B. (1991). *Imagined Communities: Reflections on the Origins and Spread of Nationalism.* London: Verso.
Apple, M. W. and Beane, J. A. (1995). "Lessons from Democratic Schools." In *Democratic schools.* Ed. M. W. Apple and J. A. Beane. Alexandria, VA: Association for Supervision and Curriculum Development.
Applegate, P. J. (2008). The Qualities That Differentiate High-Achieving and Low-Achieving High-Poverty Rural High Schools: A Transformative Mixed Methods Study. Unpublished doctoral dissertation. University of Oklahoma: Norman, Oklahoma.
Ashcroft, B., Griffiths, G. and Tuffin, H. (2000). *Post-Colonial Studies: The Key Concepts.* Routledge: London.
Ashley, M. (2005). "Can KS2 Teachers Know Enough? A Comparative Study." *Primary Practice* 40: 36–40.
Astin, A. W. and Sax, L. J. (1998). "How Undergraduates are Affected by Service Participation." *Journal of College Student Development* 39/3: 251–263.

Astley, K. and Jackson, P. (2000). "Doubts on Spirituality: Interpreting Waldorf Ritual." *International Journal of Children's Spirituality* 5/2: 221–228.
Atkinson, L. C. (2005). "Schools as Learning Organizations: Relationships between Professional Learning Communities and Technology-Enriched Learning Environments." *Dissertations Abstract International* 66/02: 466, AAT 3163444.
Averso, R. A. (2004). "A Phenomenological Study of Leadership: Developing a Shared School Vision." *Dissertation Abstracts International* 65/8, 2842.
Axtmann, R. (2001). "Introduction II: Between Polycentricity and Globalization: Democratic Governance in Europe." In *Balancing Democracy*. Ed. R. Axtmann. London: Continuum.
Ayers, W. (2003). *On the Side of the School. Summerhill Revisited*. New York: Teachers College Press.
Badawi, Hoda (2005). Parental Reasons for School Choice: A Case Study of an Islamic School in the U.S.A. PhD. dissertation. University of Minnesota Twin Cities.
Baldwin, J. M. (1906). *Thought and things: A study of the Development and Meaning of Thought* (Volume I). New York: Macmillan.
Ball, S. J. (1995). *Education Reform*. Buckingham: Open University Press.
———. (2006). *Education Policy and Social Class: The Selected Works of Stephen J. Ball*. London: Routledge.
———. (2007). *Education Plc*. London: Routledge.
Ballantyne, Y. (2006). "Teaching Māori About Asia: Print Culture and Community Identity in Nineteenth Century New Zealand." In *Asia in the Making of New Zealand*. Ed. H. Johnson, H. and B. Moloughney. Auckland: Auckland University Press.
Banks. J. (1997). *Handbook of Research on Multicultural Education*. San Francisco: Jossey-Bass/Wiley.
Barnard, C. (1999). *Two Weeks in May 1945: Sandbostel Concentration Camp and the Friends Ambulance Unit*. London: Quaker Books.
Barrington, J. (1988). "Learning the 'Dignity of Labour': Secondary Education Policy for Maoris." *New Zealand Journal of Education Studies* 20/2: 151–164.
Bar-Tal, D. (1999). "The Arab–Israeli Conflict as an Intractable Conflict and Its Reflection in Israeli Textbooks." *Megamot* 29/4: 445–491 (Hebrew).
Barth, R. S. (2005). "Turning Book Burners into Lifelong Learners." In *On common Ground: The Power of Professional Learning Communities*. Ed. R. Dufour, R. Eaker, and R. Dufour. Bloomington, IN: National Educational Service.
Batchelor, S. (1998). *Buddhism Without Beliefs: A Contemporary Guide to Awakening*. London, Bloomsbury
Battiste, M. (1998). "Enabling the Autumn Seed. Toward a Decolonized Approach to Aboriginal Knowledge, Language and Education." *Canadian Journal of Native Education* 22/1: 16–27.
———. (Ed.) (2000). *Reclaiming Indigenous Voice and Vision*. Vancouver: UBC Press.
Battiste, M. and J. Barman (Eds.) (1995). *First Nations Education in Canada: The Circle Unfolds*. Vancouver: UBC Press.
Battiste, M. and Henderson (Sa'ke'j), J. Y. (2000). *Protecting Indigenous Knowledge and Heritage: A Global Challenge*. Saskatoon, SK: Purich Publishing.
Bauman, Z. (1976). *Socialism: The Active Utopia*. London: George Allen.
———. (2000). *Liquid Modernity*. London: Polity Press.

BBC (2007). Religion and Ethics www.bbc.co.uk/religion/religions/christianity/subdivisiond/quakers_print.html. Accessed May 7, 2007.
Beaglehole, T. H. (1970). "The Missionary Schools, 1816–1840." In *Introductions to Māori Education*. Ed. J.L. Ewing and J. Shallcrass. Wellington: Price Milburn.
Beck, U. (1987). "The Anthropological Shock: Chernobyl and the Contours of the Risk Society." *Berkley Journal of Sociology* 3: 153–165.
———. (1992). *Risk Society: Towards a New Modernity*. London: Sage.
Becker, K. (2007). "Digital Game-Based Learning Once Removed: Teaching Teachers." *British Journal of Educational Technology* 38/3: 478–488.
Beeson, E. and Strange, M. (2003). *Why Rural Matters 2003: The Continuing Need for Every State to Take Action on Rural Education*. Randolph, VT: Rural School and Community Trust Policy Program.
Bekerman, Z. (2002). "Can Education Contribute to Coexistence and Reconciliation? Religious and National Ceremonies in Bilingual Palestinian–Jewish Schools in Israel." *Peace and Conflict: Journal of Peace Psychology* 8/3: 259–276.
———. (2003a) "Never Free of Suspicion." *Cultural Studies and Critical Methodologies* 3/2: 136–147.
———. (2003b) "Reshaping Conflict through School Ceremonial Events in Israeli Palestinian–Jewish Co-Education." *Anthropology and Education Quarterly* 34/2: 205–224.
———. (2004). "Multicultural Approaches and Options in Conflict Ridden Areas: Bilingual Palestinian–Jewish Education in Israel." *Teachers College Record* 106/3: 574–610.
———. (2005). "Complex Contexts and Ideologies: Bilingual Education in Conflict-Ridden Areas." *Journal of Language Identity and Education* 4/1: 1–20.
Bekerman, Z. and Horenczyk, G. (2004). "Arab–Jewish Bilingual Coeducation in Israel: A Long-Term Approach to Intergroup Conflict Resolution." *Journal of Social Issues* 60/2: 389–404.
———. (In press). "Arab–Jewish Bilingual Co-education in Israel: A Long-term Approach to Inter-group Conflict Resolution." *Journal of Social Issues*.
Bekerman, Z. and Maoz, I. (2005). "Troubles with Identity: Obstacles to Coexistence Education in Conflict Ridden Societies." *Identity* 5/4: 341–358.
Bekerman, Z. and Nir, A. (2006). "Opportunities and Challenges of Integrated Education in Conflict Ridden Societies: The Case of Palestinian–Jewish Schools in Israel." *Childhood Education* 82/6: 324–333.
Bekerman, Z. and Shhadi, N. (2003). "Palestinian Jewish Bilingual Education in Israel: Its Influence on School Students." *Journal of Multilingual and Multicultural Development* 24/6: 473–484.
Bell, D. A. (2001). "Dissenting." In *What Brown v. Board of Education Should have Said: The Nation's Top Legal Experts Rewrite America's Landmark Civil Rights Decision*. Ed. J. M. Balkin. New York: New York University Press.
Ben-Amos, A. and Bet-El, I. (1999). "Holocaust Day and Memorial Day in Israeli Schools: Ceremonies, Education and History." *Israel Studies* 4/1: 258–284.
Benton, R. (1981). *The Flight of the Amokura: Oceanic Languages and Formal Education in the South Pacific*. New Zealand Council for Educational Research: Wellington.

Berger, P. L. (1977). *Pyramids of Sacrifice: Political Ethics and Social Change*. Harmondsworth: Penguin.
Bevan-Brown, J. (2003). *The Cultural Self-Review: Providing Culturally Effective, Inclusive Education for Māori Learners*. Wellington: New Zealand Council for Educational Research.
Bhabha, H. K. (1994). *The Location of Culture*. London; New York: Routledge.
Bishop, R. and Glynn, T. (2003). *Culture Counts: Changing Power Relations in Education*. New York: Zed Books.
Blumenfeld, P. C., Soloway, E., Marx, R. W., Krajcik, J. S., Guzdial, M., and Palincsar, A. (1991). "Motivating Project-Based Learning." *Educational Psychologist* 26/3–4: 369–398.
Bolton, G. (2005). *Reflective Practice: Writing and Professional Development*. London: Sage.
Booth, J. M. and Hunn, J. K. (1962). *Integration of Māori and Pākeha*. Wellington: Government Printer.
Boston, J. Martin, J., Pallot, J., and Walsh, P. (1996). *Public Management: The New Zealand Model*. Auckland: Oxford University Press.
Bowen, W. G. and Bok, D. (1998). *The Shape of the River*. Princeton, NJ: Princeton University Press.
Boyd, W. L. (2000). "Editorial: What Counts as Educational Research?" *British Journal of Educational Studies* 48/4: 347–351.
Bransford, J., Vye, N., Bateman, H., Brophy, S. and Roselli, B. (2004). "Vanderbilt's AMIGO Project: Knowledge of How People Learn Enters Cyberspace." In *Learner-Centered Theory and Practice in Distance Education*. Ed. T. M. Duffy and J. R. Kirkley. Mahwah, NJ: Erlbaum.
Bransford, J. D., Brown, A. L., and Cocking, R. R. (Eds.) (2000). *How People Learn: Brain, Mind, Experience, and School Committee on Developments of Science of Learning*. Washington DC: National Academies Press.
Brodie, W. (1845). *Remarks on the Past and Present State of New Zealand*. London: Whittaker.
Brown v. Board of Education (1954). 347 U.S. 483 1954.
Budge, K. (2006). "Rural Leaders, Rural Places: Problem, Privilege, and Possibility." [Electronic version]. *Journal of Research in Rural Education* 21/13: 1–10.
Burnet, G. B. (2007). *The Story of Quakerism in Scotland 1650–1950*. Cambridge: Lutterworth Press.
Byers, P., Dillard, C., Easton, F., Henry, M., Mcdermott, R., Oberman, I., and Uhrmacher, B. (1996). *Waldorf Education in an Inner-City Public School: The Urban Waldorf School of Milwaukee*. Spring Valley, NY: Parker Courtney Press.
Carey, P. (Ed.) (1992). *The Elephant's Footprint: Buddhism and Education in the UK – The Dhamma School Project*. Unpublished report available from the editor.
Carlgren, F. (1972). *Education Towards Freedom: A Survey of the Work of Waldorf Schools Throughout the World*. East Grinstead: Lanthorn Press.
Casey, C. (2002). *Critical Analysis of Organizations: Theory, Practice, Revitalization*. London: Sage.
Castellano, M. B. (2000). "Updating Aboriginal Traditions of Knowledge." In *Situating Indigenous Knowledges: Definitions and Boundaries*. Ed. G. J. Sefi Dei, B. L. Hall, and D. G. Rosenberg. Toronto, ON: University of Toronto Press.

Castellano, M. B., Davis, L., and Lahache, L. (Eds.) (2000). *Aboriginal Education: Fulfilling the Promise.* Vancouver, BC: UBC Press.
Castells, M. (1998). *The Information Age: Economy, Society and Culture.* Malden, MA; Oxford: Blackwell Publishers.
Cate, J. and O'Hair, M. (2007). "K20 – ACTS Leadership for Technology Implementation: A Statewide Professional Development Model." *OASCD Journal* 14/1: 14–18.
Cate, J., Vaughn, C., and O'Hair, M. (2006). "A 17-Year Case Study of an Elementary School's Journey: From Traditional School to Learning Community to Democratic School Community." *Journal of School Leadership* 16/1: 86–111.
Chamberlin, R. (1989). *Free Children and Democratic Schools. A Philosophical Study of Liberty and Education.* New York: Falmer.
Claire, John. (2005). "Spirit of Flashman Stalks Corridors of Top Boarding Schools." *Telegraph,* June 18.
Claxton, G. (1997). *Hare Brain, Tortoise Mind: Why Intelligence Increases When You Think Less.* London: Fourth Estate.
Commission for Social Care Inspection (CSCI) (2005). *Boarding School. Summerhill School,* London, January.
Conant, J. B. (1959). *The American High School Today: A First Report to Interested Citizens.* New York: McGraw-Hill.
Council of Ministers of Education of Canada (2005). Backgrounder on the CMEC Aboriginal Education Action Plan, 1. http://www.cmec.ca/publications/index.en.stm. Accessed April 24, 2008.
Courpasson, D. and Clegg, S. (2006). "Dissolving the Iron Cages? Tocqueville, Michels, Bureaucracy and the Perpetuation of Elite Power." *Organization* 13/3: 319–343.
Court, M. R. (2003). "Towards Democratic Leadership. Co-principal Initiatives." *International Journal of Leadership in Education* 6/2: 161–183.
———. (2004). "Talking back to New Public Management Versions of Accountability in Education. A Co-principalship's Practices of Mutual Responsibility." *Educational Management, Administration and Leadership* 32/2: 173–195.
———. (2005). "Crossing Cultural Boundaries or Caught in Political Crossfire?" *New Zealand Journal of Educational Leadership* 20/1: 47–64.
Court, M. R. and Waitere-Ang, H. J. (2004). Nau te raurau, naku te raurau, ka ora ai te kura: Building Bi-cultural Leadership in Schools. A paper presented in the Symposium, "Turning the co-principalship kaleidoscope: past, present and future developments" at the New Zealand Association of Research in Education Annual Conference Turning the Kaleidoscope, at Wellington Westpac Stadium, November 26.
Cox, L. (1993). *Kotahitanga; The Search For Maori Political Unity.* Oxford University Press: Auckland.
Cox, M. and Rolands, A. (2000). "The Effect of Three Different Educational Approaches on Children's Drawing Ability: Steiner, Montessori and Traditional." *British Journal of Educational Psychology* 70: 485–503.
Crain, W. (2000). *Theories of development: Concepts and Applications.* Upper Saddle River, NJ: Prentice Hall.
Csikszentmihalyi, M. (1990). *Flow.* New York: Harper & Row.

———. (1996). *Creativity: Flow and the Psychology of Discovery and Invention*. New York: Harper Collins.
Csikszentmihalyi, M. and Schneider, B. (2000). *Becoming Adult: How Teenagers Prepare for the World of Work*. New York: Basic Books.
Csikszentmihalyi, M., Rathunde, K., and Whalen, S. (1997). *Talented Teenagers: The Roots of Success and Failure*. New York: Cambridge University Press.
Cuban, L. (2000). "Why Is It So Hard to Get Good schools?" In *Reconstructing the Common Good in Education: Coping with Intractable American Dilemmas*. Ed. L. Cuban and D. Shipps. Stanford, CA: Stanford University Press.
Curie, A. (2007). *True Justice, Quaker Peace Makers and Peace Making*. Quaker Peace and Social Witness.
Dahl, R. (1994). "After the Revolution? The Authority in Advanced Societies." Barcelona: Gedisa Editorial.
Damasio, A. (1994). *Decartes' Error*. New York: Penguin.
Dandelion, P. (1996). *A Sociological Analysis of the Theology of Quakers: The Silent Revolution*. Ceredigion: Edward Mellen Press.
Darling-Hammond, L. (1997). *The Right to Learn: A Blueprint for Creating Schools That Work*. San Francisco: Jossey-Bass Publishers.
———. (2000). "Teacher Quality and Student Achievement: A Review of State Policy Evidence." *Education Policy Analysis Archives* 8/1. Retrieved from http://epaa.asu.edu/epaa/v8n1/. Accessed May 19, 2004.
Davies, L. and Nicholl, K. (1993). *Te Māori i Roto i Ngā Mahi Whakākoranga: Māori in Education*. Wellington: Ministry of Education.
Dávila, A. and Mora, M. T. (2007). *Civic Engagement and High School Academic Progress: An Analysis Using NELS Data. Circle Working Paper 52*. College Park, MD: The Center for Information and Research on Civic Learning and Engagement. Available online at www.civicyouth.org.
Dawkins, R. (2006). *The God Delusion*. London, Bantam Press.
De Oliveira, B. M. (2002). O Colegiado na escola Pública: Uma experiencia compartilhada no exercício da cidadania? Um estudo da caso junto ao Colegiado do Colegio Estadual "Gobernador Milton Campos."—Belo Horizonte—MG. Unpublished master thesis: Pontificia Universidade Católica de Minas Gerais.
De Ruyter, D. and Merry, M. S. (In press). "Why Education in Public Schools Should Include Religious Ideals," *Studies in Philosophy and Education*.
Deal, T. and Kennedy, A. (1982). *Corporate Cultures*. Reading, MA: Addison-Wesley Publishing Co.
Deci, E. L. and Ryan, R.M. (1985). *Intrinsic Motivation and Self-Determination in Human Behavior*. New York: Plenum.
Department of Education. (1987). *The Curriculum Review, Report of the Committee to Review the Curriculum for Schools*. Wellington: Government Printer.
———. (1988). *Tomorrow's Schools: The Reform of Education Administration in New Zealand*. Wellington: Government Printer.
Department for Education and Employment (DfEE) (1994). *Our Children's Education*. London: DfEE.
Dewey, J. (1910). *How We Think*. Mineola, New York: Dover.
———. (1916). *Democracy and Education: An Introduction to the Philosophy of Education*. New York: The MacMillan Company.
———. (1934/1980). *Art as Experience*. New York: Perigee.

———. (1962). *The Child and the Curriculum, and, The School and Society*. Chicago: Chicago University Press (first published 1902, 1900, respectively).

———. (1966). *Democracy and Education. An Introduction to the Philosophy of Education*. New York: The Free Press.

Dickason, O. P. (1992). *Canada's First Nations: A History of the Founding Peoples from Earliest Times*. Toronto: Oxford University Press.

Dillon, J. (2000). "The Spiritual Child: Appreciating Children's Transformative Effects on Adults." *Encounter: Education for Meaning and Social Justice* 13/4: 4–18.

Dion, S. (2008). *Braiding Histories: Learning from Aboriginal Peoples' Experiences and Perspectives*. Vancouver: UBC Press.

Donaldson, M. (1978). *Children's Minds*. Glasgow: Fontana.

Dow, S. (2006). Collaborating with Parents to Enhance the Effectiveness of a Bicultural School Learning Community. A research project presented in partial fulfilment of the requirements for the degree of Master of Educational Administration at Massey University, Palmerston North, New Zealand.

Driessen, G. (2007). *Opbrengsten van Islamitische Basisscholen. Prestaties, Attitudes en Gedrag van Leerlingen op Islamitische Scholen Vergeleken* (Attainment of Islamic Primary Schools. Achievement, Attitudes and Behavior of Pupils at Islamic Schools Compared). Nijmegen: ITS.

———. (2008). "De Verwachtingen Waargemaakt? Twee Decennia Islamitische Basisscholen." *Mens & Maatschappij* 83: 168–189 (Expectations Met? Two Decades of Islamic Primary Schools).

Driessen, G. and Bezemer, J. (1999). "Background and Achievement Levels of Islamic Schools in the Netherlands: Are the Reservations Justified?" *Race Ethnicity and Education* 2: 235–256.

Driessen, G. and Merry, M. S. (2006). "Islamic Schools in the Netherlands: Expansion or Marginalization?" *Interchange* 37: 201–223.

Driessen, G. and Valkenberg, P. (2000). "Islamic Schools in the Netherlands: Compromising between Identity and Quality?" *British Journal of Religious Education* 23: 15–26.

Dryzek, J. S. (1996). *Democracy in Capitalist Times*. Oxford: Oxford University Press.

Dufour, R., Eaker, R., and Dufour, R. (2002). *Getting Started: Reculturing Schools to Become Professional Learning Communities*. Bloomington, IN: National Educational Service.

Dumasy, E. (2008). *Kwaliteitsdilemma's van Islamitische Scholen* (Quality dilemmas of Islamic Schools). Amsterdam: Uitgeverij SWP.

Durie, M. (1993). "Māori and the State: Professional and Ethical Implications for the Public Service." *Proceedings of the Public Service Senior Management Conference*, Wellington, 22–35.

———. (1997). Identity, Access And Maori Advancement. Keynote address at New Zealand Educational Administration Society Research Conference, *New Directions in Educational Leadership: The Indigenous Future*. 6–8th July. Auckland Institute of Technology: Auckland.

Dwyer, C. and Meyer, A. (1995). "The Institutionalisation of Islam in the Netherlands and in the UK: The Case of Islamic Schools." *New Community* 21: 37–54.

Dyck, J. (2002). "The Built Environment's Effect on Learning: Applying Current Research." *Montessori Life* 14: 53–56.
Easton, F. (1997). Educating the Whole Child, "head, heart and hands": Learning from the Waldorf experience, *Theory into Practice*, 36 (2) 87–95.
Eccles, J., Midgley, C., Wigfield, A., Buchanan, C. M., Reuman, D., Flanagan, C., et al. (1993). "Development During Adolescence: The Impact of Stage-Environment Fit on Young Adolescents' Experience in Schools and Families." *American Psychologist* 48: 90–101.
Education Trust. (2003). *Education watch: The nation.* http://www2.edtrust.org/edtrust/summaries2004/USA.pdf. Accessed May 8, 2006.
Elsevier (2008). Moslimschool Moet Gestolen Miljoen Terugbetalen (Muslim School Has To Pay Back Stolen Millions). http://www.elsevier.nl/web/nederland/moslimschoolmoetgestolenmiljoenterugbetalen.htm. Accessed August 27, 2008.
Erdrich, L. (2004). *Four Souls.* New York: Harper Collins Publishers.
Ermine, W. (1995). "Aboriginal Epistemology." In *First Nations Education in Canada: The Circle Unfolds.* Ed. M. Battiste and J. Barman. Vancouver, BC: UBC Press.
Eseryel, D. (2006). "Expert Conceptualizations of the Domain of Instructional Design: An Investigative Study on the DEEP Assessment Methodology for Complex Problem-Solving Outcomes." Unpublished Ph.D. Dissertation. Syracuse University, Syracuse, NY.
Feola, M. (1997). *George Bishop: Seventeenth-Century Soldier Turned Quaker.* London: William Sessions Ltd.
Feueruerger, G. (1998). "Neve Shalom/Wahat Al-Salam: A Jewish-Arab School for Peace." *Teachers College Record* 99: 692–730.
———. (2001). *Oasis of Dreams: Teaching and Learning Peace in a Jewish–Palestinian Village in Israel.* New York: RoutledgeFalmer.
Fielding, M., Bragg, S., Craig, J., Cunningham, I., Eraut, M., Gillinson, S., Horne, M., Robinson, C., and Thorp, J. (2005). *Factors Influencing the Transfer of Good Practice.* London: Department for Education and Skills.
Foreman, C. W. (1996). Service Learning in the Small Group Communication Class. Paper presented at the Annual Meeting of the Speech Communication Association, San Diego, CA.
Foucault, M. (1977). *Discipline and Punish: The Birth of the Prison.* Trans. A. Sheridan. Penguin: Harmondsworth.
Fouts and Associates (2001). The Bill and Melinda Gates Foundation's Model School Initiative Year 1 Evaluation. Accessed May 5, 2006 from http://www.gatesfoundation.org/ nr/Downloads/ed/researchevaluation/SchoolsFinalReport2001.pdf.
Freire, P. (1970). *Pedagogy of the Oppressed.* New York: Continuum.
Fullan, M. (1991). *The New Meaning of Educational Change.* New York: Teachers College Press.
———. (2001a). *Leading in a Culture of Change.* San Francisco: Jossey-Bass.
———. (2001b). *The New Meaning of Educational Change.* Third ed. New York: Teachers College Press.
———. (2003). *Change Forces with a Vengeance.* New York: Routledge.
Fundaçao de Economia e Estatistica (2005). Secretaria de Coordenaçao e Planejamento – Governo do Estado do Rio Grande do Sul. Available online at http://www.fee.rs.gov.br.

Furman, G. C. and Gruenwald, D. A. (2004). "Expanding the Landscape of Social Justice: A Critical Ecological Analysis." *Educational Administration Quarterly* 40/1: 47–76.
Furman, G. C. and Shields, C. M. (2003). How Can Educational Leaders Promote and Support Social Justice and Democratic Community in Schools?. A paper presented at the annual meeting of the American Educational Research Association in Chicago, IL. Available online at http://www.aera.net/divisions/a/taskforce/ Furman%20Shields4-28.pdf. Gallagher, N. (2007). *Quakers in the Israeli–Palestinian Conflict: The Dilemmas of NGO Humanitarian Activism.* The American University in Cairo Press.
Gallagher, S. (2005). *How the Body Shapes the Mind.* New York: Oxford University Press.
Gallagher, T. (2007). "Desegregation and Resegragation: The Legacy of Brown versus Board of Education, 1954." In *Addressing Ethnic Conflict Through Peace Education: International Perspectives.* Ed. Z. Bekerman and C. McGlynn. New York: Palgrave Macmillan.
Gandin and Apple (2002). "Challenging Neoliberalism, Building Democracy: Creating the Citizen School in Porto Alegre, Brazil." *Journal of Education Policy* 17/2: 259–279.
Garcia, O. (1997). "Bilingual Education." In *The Handbook of Sociolinguistics.* Ed. F. Coulmas. Oxford: Blackwell.
Garet, M. S., Porter, A. C., Desimore, L. M., Birman, B.T., and Yoon, K. S. (2001). "What Makes Professional Development Effective? Results from a National Sample of Teachers." *American Educational Research Journal* 38/4: 915– 945.
Gellner, E. (1997). *Nationalism.* New York: New York University Press.
Ghanem, A. (1998). "State and Minority in Israel: The Case of Ethnic State and the Predicament of Its Minority." *Ethnic and Racial Studies* 21/3: 428–448.
Gilligan, C. (1982). *In a Different Voice: Psychological Theory and Women's Development.* Cambridge, MA: Harvard University Press.
Gillman, H. (1988). *A Light That is Shining: Introduction to the Quakers.* First edition. London: Quaker Home Service.
———. (2003). *A Light That is Shining: Introduction to the Quakers.* Third edition. London: Quaker Books.
Giroux, H. (1992). *Border Crossings. Cultural Workers and the Politics of Education.* New York: Routledge.
Glazier, J. A. (2003). "Developing Cultural Fluency: Arab and Jewish Students Engaging in One Another's Company." *Harvard Educational Review* 73/2: 141–163.
———. (2004). "Collaborating with the "Other": Arab and Jewish Teachers Teaching in Each Other's Company." *Teachers College Record* 106/3: 611–633.
Glickman, C. D. (1993). *Renewing America's Schools: A Guide for School-Based Action.* San Francisco: Jossey-Bass.
———. (1998). *Revolutionizing America's schools.* San Francisco: Jossey-Bass.
Glickman, C. D. and Alridge, D. P. (2003). "Going Public: The Imperative of Public Education in the Twenty-first century." In *Renewing America's schools: A guide for School-Based Action.* C. D. Glickman. San Francisco: Jossey- Bass.
Glickstein, H. A. (1996). "Inequities in Educational Funding." In *Brown v. Board of Education: The Challenge for Today's Schools. New York.* Ed. E. C. Lagemann and L. P. Miller. New York: Teachers college Press.

Goldacre, B. (2006). "Bad Science," *The Guardian*, Saturday, March 18.
Golden, J. (1997). Narrative - The Use of Story in Waldorf Education. Paper presented at Annual Meeting of American Educational Research Association, Chicago.
Good, T. L. and Brophy, J. E. (2000). *Looking in Classrooms*. Eighth edition. New York: Longman.
Gorman, G. (1973). *The Amazing Fact of Quaker Worship*. London: Quaker Home Service.
Goswami, U. (2006). "Neuroscience and Education: From Research to Practice?" *Nature Reviews Neuroscience* 7: 406–413. Available online at www.nature.com/nrn/index.html.
Graham, J. P. H. (2002). Nāu te rūnanga, nāku te rūnanga, ka piki ake te ōranga o te iwi : Partnership Relationships between Schools and Māori Communities. A thesis presented in partial fulfilment of the requirements for the degree of Master of Education at Massey University, Palmerston North, New Zealand.
Grande, S. (2004). *Red Pedagogy: Native American Social and Political Thought*. New York: Rowman and Littlefield.
Graveline, F. J. (2003). *Circle Works: Transforming Eurocentric Consciousness*. Blackpoint, NS: Fernwod Books.
Greene, B. A., Miller, R. B., Crowson, H. M., Duke, B. L., and Akey, C. L. (2004). "Relations among Student Perceptions of Classroom Structures, Perceived Ability, Achievement Goals, and Cognitive Engagement and Achievement in High School Language Arts." *Contemporary Educational Psychology* 29/4: 462–482.
Greeno, J. G. (1997). "On Claims That Answer the Wrong Questions." *Educational Researcher* 26/1: 5–17.
Gruenewald, D. A. (2003). "Foundations of Place: A Multidisciplinary Framework for Place-Conscious Education." [Electronic version] *American Educational Research Journal* 40/3: 619–654.
Guinier, L. and Torres, G. (2002). *The Miner's Canary: Enlisting Race, Resisting Power, and Transforming Democracy*. Cambridge, MA: Harvard University Press.
Gutman, A. (1993). "Democracy and Democratic Education." [Electronic version]. *Studies in Philosophy and Education* 12: 1–9.
Haig-Brown, C. (1988). *Resistance and Renewal: Surviving the Indian Residential School*. Vancouver: Arsenal Pulp Press.
———. (1995). *Taking Control: Power and Contradiction in First Nations Adult Education*. Vancouver, BC: UBC Press.
Haig-Brown, C. and Dannenmann, K. (2002). "A Pedagogy of the Land: Dreams of Respectful Relations." *McGill Journal of Education* 37/3: 451–468.
Halpin, D. (2003a). *Hope and Education: The Role of the Utopian Imagination*. London, Routledge Falmer.
———. (2003b). "Hope, Utopianism and Educational Renewal." *The Encyclodeia of Informal Education*. www.infed.org/biblio/hope.htm. Last updated January 30, 2005. (A fuller version of the paper was present at Charterhouse School, Monday, January 6, 2003.) Accessed May 7, 2007.
Halstead, J. M. (1993). "The Case for Single-sex Schools: A Muslim Approach." *Muslim Education Quarterly* 10: 49–69.
———. (1995). "Voluntary Apartheid? Problems of Schooling for Religious and Other Minorities in Democratic Societies." *Journal of Philosophy of Education* 29/2: 257–272.

———. (2007). "In Place of a Conclusion: The Common School and the Melting Pot." *Journal of Philosophy of Education* 41/4: 829–842.
Hamlin, G. (2007). *An Evaluation Report to the Oklahoma Educational Technology Trust Concerning Sustainability of Grants to Schools.* Oklahoma City, OK: OETT.
Hampshire Quakers Web site (www.hampshirequakers.org.uk/quakers/testimonies.htm) Quaker Testimonies. Accessed February 28, 2007.
Hampton, E. (1995). "Towards a Redefinition of Indian Education." In *First Nations Education in Canada: The Circle Unfolds.* Ed. M. Battiste and J. Barman. Vancouver, BC: UBC Press.
Handelman, D. (1990). *Models and Mirrors: Towards an Anthropology of Public Events.* Cambridge: Cambridge University Press.
Hannafin, M. J., and Land, S. M. (1997). "The Foundations and Assumptions of Technology-Enhanced, Student-Centered Learning Environments." *Instructional Science* 25/3: 167–202.
Hardy, A. (1966). *The Divine Flame: Natural History and Religion.* London: Collins.
———. (1979/1991). *The Spiritual Nature of Man.* Oxford: The Alister Hardy Research Centre.
Hargreaves, A. and Fink, D. (2006). *Sustainable Leadership.* San Francisco, CA: Jossey Bass.
Harker, R. (1985). "Schooling and Cultural Reproduction." In *Political issues in New Zealand Education.* Ed. J. Codd, R. Harker, and R. Nash. Palmerston North: Dunmore Press.
Harker, R. K. and McConnochie, K. R. M. (1987). *Education as a Cultural Artifact. Studies in Māori and Aboriginal Education.* Palmerston North: Dunmore Press.
Hatton, J. (2005). *Betsy:Tthe Dramatic Biography of Prison-Reformer Elizabeth Fry,* London: Monarch.
Haw, Karen F. (1994). "Muslim Girls' Schools: A Conflict of Interests?" *Gender and Education* 6: 63–76.
Hay, D. (1987). *Exploring Inner Space.* Second edition. Harmondsworth: Penguin.
Hay, D. and Hunt, K. (2000). *Understanding the Spirituality of People Who Don't Go to Church.* Nottingham: University of Nottingham.
Haydon, G. (2007). "In Search of the Comprehensive Ideal: By Way of an Introduction." *Journal of Philosophy of Education* 29/2: 523–538.
Henderson, J. G. and Slattery, P. (2006). "Editors' Introduction: Democracy, Culture, and Spirituality in the International Arena." *Journal of Curriculum and Pedagogy* 2/2: 1–7.
Henry, M. (1992). "School Rituals as Educational Contexts: Symbolising the World, Others and Self in Waldorf and College Pre Schools." *International Journal of Qualitative Studies in Education* 5/4: 295–309.
Her Majesty's Inspectorate (HMI)/Office for Standards in Education (OfSTED) (1999). *Report on Summerhill School.* OfSTED, March.
Hermans, P. (1995). "Moroccan Immigrants and School Success." *International Journal of Educational Research* 23: 33–43.
Heron, A. (1997). *The British Quakers 1647–1997: Highlights of Their History.* Kelso, Scotland: Curlew Productions.

Heron, A. (2000). *On Being a Quaker: Membership Past-Present-Future*. Kelso Scotland: Curlew Productions.
Hesketh, A. and Knight, P. (1998). "Secondary School Prospectuses and the Educational Market." *Cambridge Journal of Education* 28/1: 21–35.
Hindess, B. (2001). "Democracy, Multiculturalism and the Politics of Difference." In *Balancing Democracy*. Ed. R. Axtmann. London: Continuum.
Hirschman, A. (1970). *Exit, Voice and Loyalty. Responses to Decline in Firms, Organizations and States*. Cambridge: Harvard University Press.
Hodson, J. H. (2004). Learning and Healing: A Wellness Pedagogy for Aboriginal Teacher Education. Unpublished master's thesis, Brock University, St. Catharines, Ontario, Canada.
Holden, C. and Clough, N. (2002). *Education for Citizenship: Ideas into Action – A Practical Guide for Teachers*. London: Routledge.
Hord, S. M. (1997). *Professional Learning Communities: Communities of Continuous Inquiry and Improvement*. Austin, TX: Southwest Educational Development Laboratory.
House, R. (2001). "Stress and the Waldorf Teacher: Towards Pre-emption Through Understanding." *Steiner Education* 35/2: 36–41.
Huffman, J. B. and Hipp, K. K. (2003). *Reculturing Schools as Professional Learning Communities*. Lanham, MD: Scarecrow Education.
Hunn, J. K. (1960). *Report on Department of Māori Affairs*. Wellington: Government Printer.
Husain, E. (2007). *The Islamist*. London: Penguin Books.
Huxham, C. (2003). "Theorizing Collaboration Practice." *Public Management Review* 5/3: 401–423.
IHFHR. (2005). *Intolerance and Discrimination against Muslims in the EU: Developments since September 11*, Vienna: International Helsinki Federation for Human Rights.
Inal, Y. and Cagiltay, K. (2007). "Flow Experiences of Children in an Interactive Social Game Environment." *British Journal of Educational Technology* 38/3: 455–464.
Independent Schools Inspectorate. (2004). Leighton Park School. www.isinspect.org.uk/reports/2004/0641_04_r.htm. Accessed February 28, 2007.
Instituto Brasileiro de Geografia e Estatística (2006). *Ensino - matrículas, docentes e rede escolar 2006*. Available Online at www.ibge.gov.br.
Irwin, K. (1988). "Racism and Education." In *Getting it Right. Aspects of Ethnicity and Equity in New Zealand Education*. Ed. W. Hirsh and R. Scott. Wellington: Office of the Race Relations Conciliator.
Istanbouli, Mazen (2000). An Exploratory Case Study of Religio-Cultural Issues in an Islamic School: Implications for Socialization and Enculturation. Ph.D. dissertation, Loyola University, Chicago.
Jackson, M. (1998). Research and the Colonisation of Maori knowledge. Paper presented at Oru Rangahau: Maori Research and Development conference: Massey University.
———. (2007). "Globalisation and the Colonising State of Mind." In *An Indigenous Response to Neoliberalism*. Ed. M. Bargh. Huia Publishers: Wellington.
James, W. (1985 [1902]). *The Varieties of Religious Experience*. Harmondsworth: Penguin Books.

Jawad, H. and Benn, T. (2003). *Muslim Women in the United Kingdom and Beyond: Experiences and Images.* Leiden: Brill.
Jelinek, D. and Sun, L. (2003). *Does Waldorf Offer a Viable Form of Science Education?* Sacramento, CA: College of Education, California State University.
Jenkins, K. and Ka'ai, T. (1994). "Maori Education: A Cultural Experience and Dilemma for the State - A New Direction for Maori Society." In *The Politics of Learning and Teaching in Aotearoa - New Zealand.* Ed. E. Coxen, K. Jenkins, J. Marshall, and L. Massey. Palmerston North: Dunmore Press.
Jimerson, L. (2003). *The Competitive Disadvantage: Teacher Compensation in Rural America.* Washington, DC: Rural School and Community Trust.
Johnson, J. and Strange, M. (2007). *Why Rural Matters 2007: The Realities of Rural Education Growth.* Arlington, VA: Rural School and Community Trust.
Johnson, M. (1987). *The Body in the Mind: The Bodily Basis of Meaning, Imagination, and Reason.* Chicago: The University of Chicago Press.
Johnston, P. (1993). Examining a State Relationship: Legitimation and Te Kohanga Reo – the Return of the Prodigal Child? A paper presented at the New Zealand Association of Research in Education Conference. Hamilton, University of Waikato.
Johnston, P. and McLeod J. (2001). "Disrupting Hegemonic Spaces: Challenging Teachers to Think Indigenised." In *Canadian Indigenous / Native Studies Association. Proceedings of the Annual Conference,* "In Partnership." May 31–June 3.
Jonassen, D. H. (1999). "Designing Constructivist Learning Environments." In *Instructional Design Theories and Models, Vol. 2: A new paradigm of instructional technology.* Ed. C. M. Reigeluth. Mahwah, NJ: Lawrence Erlbaum.
Joy, B. (2006). "Mentoring and Coaching." *Research Matters,* 27, The London Centre for Leadership in Learning, Institute of Education, University of London, Spring.
Kabdan, R. (1992). "Op Weg naar Maatschappelijke Spanningen op Religieuze Gronden. Islamitische Scholen in Nederland Zijn Onwenselijk" (Toward Societal Tensions on Religious Grounds: Islamic Schools in The Netherlands Are Undesirable). *Vernieuwing* 51: 4–8.
Kahn, D. (1997). "Normalization and Normality across the Planes of Development." *NAMTA Journal* 22: 122–136.
Keen, C. and Keen, J. (1998). *Bonner Student Impact Survey.* Bonner Foundation.
Keenan, B. (1997). *Spirit in the Shadows.* London: The National Society's RE Centre.
Kelly, P. (1999). "Integration and Identity in Muslim Schools: Britain, United States and Montreal." *Islam and Christian–Muslim Relations* 10/2: 197–217.
Kelsey, J. (1995). *The New Zealand Experiment: A World Model for Structural Adjustment?* Auckland: Auckland University Press with Brigid Williams Books.
Kendall, T. A (1815). *A Korao no New Zealand: Or The New Zealanders' First Book: Being an Attempt to Compose some Lessons for the Instruction of the Natives.* Sydney: publisher unknown.
Kincheloe, J. L. (1991). *Teachers as Researchers: Qualitative Inquiry as a Path to Empowerment.* London: The Falmer Press.

Kompf, M. and Hodson, J. (2000). "Keeping the Seventh Fire: Developing an Undergraduate Degree Program for Aboriginal Adult Educators." *Canadian Journal of Native Education* 24/2: 185–202.

Labouvie-Vief, G. (1994). *Pscyh & Eros: Mind and Gender in the Life Course.* New York: Cambridge University Press.

Lakeland, P. (1997). *Postmodernity; Christian Identity in a Fragmented Age.* Minneapolis: Fortress Press.

Lakoff, G. and Johnson, M. (1999). *Philosophy in the flesh.* New York: Basic Books.

Lambert, L. (1998). *Building Leadership Capacity in Schools.* Alexandria, VA: Association for Supervision and Curriculum Development.

Lamkin, M. L. (2006). "Challenges and Changes Faced by Rural Superintendents." *Rural Educator.* http://findarticles.com/p/articles/mi_qa4126/is_200610/ai_n16840815. Accessed June 5, 2007.

Lave, J. and Wenger, E. (1990). *Situated Learning: Legitimate Peripheral Participation.* Cambridge, UK: Cambridge University Press.

Law, J. and Mol, A. (2002). "Local Entanglements or Utopian Moves: An Inquiry into Train Accidents." In *Utopia and Organisation.* Ed. M. Parker. London: Blackwell/Sociological Review.

Lawson M. (Ed.) (1973). *Summerhill: For and Against. Assessments of A.S. Neill.* Sydney: Angus and Robertson.

Leder, D. (1990). *The Absent Body.* Chicago, IL: University of Chicago Press.

Lee, V. E. and Smith, J. B. (1996). "Collective Responsibility for Learning and Its Effects on Gains in Achievement for Early Secondary School Students." *American Journal of Education* 104/2: 103–147.

———. (2001). *Restructuring High Schools for Excellence: What Works.* New York: Teachers College Press.

Lester, J. (1920). *The Ideals and Objectives of Quaker Education.* Philadelphia, PA: Friends Council on Education.

Levenduski, C. (1996). *Peculiar Power: Quaker Woman Preacher in Eighteenth-Century America.* Washington, DC: Smithsonian Books.

Lewis, C. (2002). "Does Lesson Study Have a Future in the United States?" *Journal of the Nagoya University Education Department* 1/1: 1–24.

Lewis, C., Perry, R., and Hurd, J. (2004). "A Deeper Look at Lesson Study." *Educational Leadership* 61/5: 6–11.

Lillard, A. (2005). *Montessori: The Science Behind the Genius.* New York: Oxford University Press.

Lim, C. P., Nonis, D., and Hedberg, J. (2006). "Gaming in a 3D Multiuser Virtual Environment: Engaging Students in Science Lessons." *British Journal of Educational Technology* 37/2: 211–231.

Little, J. (1993). "Teachers' Professional Development in a Climate of Educational Reform." *Educational Evaluation and Policy Analysis* 15/2: 129–152.

Lloyd, S and Wernham, S. (2005). *Jolly Phonics*, Chigwell: Jolly Learning.

Loeffler, M. H. (1992). *Montessori in Contemporary American Culture.* Portsmouth, NH: Heinemann.

Long, M., (2000). *The Psychology of Education.* London: Routledge Falmer.

Loomba, A. (1998). *Colonialism/Postcolonialism.* New York: Routledge.

Louis, K., Marks, H., and Kruse, S. (1996). "Teachers' Professional Community in Restructuring Schools." *American Educational Research Journal* 33/4: 757–798.

Loukes, H. (1958). *Friends and Their Children.* London: Quaker Home Service Committee.
Lucas, H. and Lamb, A. (2000). Neill's diamonds. An oral history of Summerhill School (mimeo). Leiston, Suffolk: Summerhill School.
MacBeath, J. (2004). "Democratic Learning and School Effectiveness: Are They by any Chance Related?" In *Democratic Learning: The Challenge to School Effectiveness.* Ed. J. MacBeath and L. Moos. London: RoutledgeFalmer.
MacCulloch, D. (2003). *Reformation: Europe's House Divided 1490–1700.* London: Penguin.
McDermott, R., Byers, P., Dilliard, C., Easton, F., Henry, M. and Uhrmacher, P.B. (1996) Waldorf Education in an Inner-city Public School, *Urban Review* 28(2) 119–190.
Malhoit, G. C. (2005). *Providing Rural Students with a High Quality Education: The Rural Perspective on the Concept of Educational Adequacy.* Arlington, VA: The Rural School and Community Trust. http://www.ruraledu.org/site/apps/s/link.asp?c=beJMIZOCIrH&b=1000949. Accessed April 23, 2006.
Malhoit, G. C. and Gottoni, N. (Eds.) (2003). *The Rural School Funding Report.* Arlington, Virginia: The Rural School and Community Trust, vol. 2, 12.
Manji, I. (2003) *The Trouble with Islam Today,* New York: St. Martin's Press.
Marcel, G. (1951). *Homo Viator: Introduction to a Metaphysics of Hope.* London: Victor Gollancz.
Marks, M. (1984). "The Frustrations of Being a Māori Language Teacher." In *Report Ngā Tumanako, the Māori Educational Development Conference.* Ed. R. J. Walker. Auckland: University of Auckland.
Marples, R. (2006). "Against Faith Schools: A Philosophical Argument for Children's Rights." In *Reflecting on Faith Schools.* Ed. H. Johnson. Abingdon: Routledge.
Martindale, C. (1999). "Biological Bases of Creativity." In *Handbook of Creativity.* Ed. R. Sternberg. New York: Cambridge University Press.
Maslow, A. (1968). *Toward a Psychology of Being.* New York: Van Nostrand Reinhold.
Masters, B. (1996). An Appraisal of Steinerian Theory and Waldorf Praxis: How do They Compare? PhD edn. London: Roehampton Institute.
McGlynn, C. W. (2001). *The Impact of Post Primary Integrated Education in Northern Ireland on Past Pupils: A Study.* Belfast: University of Ulster at Jordanstown.
McLaughlin, M. W. and Talbert, J. E. (2001). *Professional Communities and the Work of High School Teaching.* Chicago: University of Chicago Press.
McLeod, J. (2002). Better Relationships for Better Learning: Schools Addressing Māori Achievement through Partnership. A thesis submitted as partial fulfilment of a Masters degree in Education at Te Uru Maurarau, Massey University College of Education, Palmerston North.
Meier, D. (2000). "Progressive Education in the 21st Century: A Work in Progress." In *Education in a New Era (2000 ASCD Yearbook).* Ed. R. S. Brandt. Alexandria, VA: Association for Supervision and Curriculum Development. Available online at http://www.ascd.org/portal/site/ascd/template.book/menuitem. 83f4b2b5537730a98d7ea23161a001ca/?bookMgmtId=8acf177a55f9ff00VgnVC M1000003d01a8c0RCRD.
———. (2002). *In Schools We Trust: Creating Communities of Learning in an Era of Testing and Standardization.* Boston: Beacon Press.

Merry, M. S. (2004). "Islam vs (Liberal) Pluralism?" *Journal of Muslim Minority Affairs* 24: 121–137.

———. (2005a). "Advocacy and Involvement: The Role of Parents in Western Islamic Schools." *Religious Education* 100: 374–385.

———. (2005b). "Should Educators Accommodate Intolerance? Mark Halstead, Homosexuality and the Islamic Case." *Journal of Moral Education* 34: 19–36.

———. (2005c). "Social Exclusion of Muslim Youth in Flemish- and French-Speaking Belgian Schools." *Comparative Education Review* 49: 1–22.

———. (2007). *Culture, Identity and Islamic Schooling: A Philosophical Approach*. New York: Palgrave.

Merry, M. S. and Driessen, G. (2005). "Islamic Schools in Three Western Countries: Policy and Procedure." *Comparative Education* 41: 411–432.

Merry, M. S. and Milligan, J. A. (In press). "Complexities of Belonging in Democratic/Democratizing Societies: Islamic Identity, Ethnicity and Citizenship in the Netherlands and Aceh." *Journal of Muslim Minority Affairs*

Merry, M. S. and New, W. (in press, 2008). "Constructing an Authentic Self: The Challenges and Promise of African-Centered Pedagogy." *American Journal of Education* 115/1: 135–64

Milloy, J. S. (1999). *A National Crime: The Canadian Government and the Residential School System, 1879 to 1986*. Winnipeg, MN: The University of Manitoba Press.

Mills, J. (2005). *Music in the School*. Oxford: Oxford University Press.

Ministry of Education Ontario, Aboriginal Education Office (2007). *Ontario First Nation, Métis, and Inuit Education Policy Framework*. Available at http://www.edu.gov.on.ca. Accessed April 24, 2008.

Minnick, E. (2000). In *Protecting Indigenous Knowledge and Heritage: A Global Challenge*. M. Battiste and J. Y. Henderson (Sa'ke'j). Saskatoon, SK: Purich Publishing.

MinOCW (2007a). *Brief aan de Voorzitter van de Tweede Kamer der Staten Generaal* (Letter to the Chairman of the House of Representatives) *September 25, 2007*. PO/BB/07/35017. Onderwerp: Islamitisch Onderwijs. Den Haag: Ministerie van Onderwijs, Cultuur en Wetenschap.

MinOCW (2007b). Raport van een Incidenteel Onderzoek naar de Islamitische Scholengemeenschap Ibn Ghaldoun te Rotterdam (Report of an Incidental Investigation of the Islamic School Community Ibn Ghaloud in Rotterdam). Den Haag: Ministerie van Onderwijs, Cultuur en Wetenschap.

Mintzberg, H. (1993). "La estructuración de las organizaciones." En El Proceso Estratégico. H. Mintzberg y J. Quinn. Prentice Hall Hispanoamericana.

Mitchell, A. and Savill-Smith, C. (2004). *The Use of Computer and Video Games for Learning: A Review of the Literature*. London: The Learning and Skills Development Agency.

Mohamed, Madhi (2005). A Comparison of the Mathematical Performance of Grade 3 Students in Provincial Public and Islamic Private Schools in Ontario, M.A. Thesis, University of Windsor.

Montessori, M. (1917). *Spontaneous Activity in Education: The Advanced Montessori Method*. Trans. F. Simmonds. New York: Schocken.

———. (1946). *Unpublished Lectures*. London: AMI.

———. (1966). *The Secret of Childhood*. Trans. M. J.Costello. New York: Balantine.

———. (1973). *From Childhood to Adolescence*. Madras: Kalakshetra Publications.
———. (1981). *The Secret of Childhood*. New York: Ballantine.
———. (1988). *The Discovery of the Child*. Oxford: Clio Press.
———. (1989). *To Educate the Human Potential*. Oxford: Clio Press.
Morrisseau, C. (1999). *Into the Daylight: A Wholistic Approach to Healing*. Toronto, ON: University of Toronto Press.
Mulford, B. (2005). "The International Context for Research in Educational Leadership." *Educational Management, Administration and Leadership* 33/2: 150.
Murphy, J. (2000). "Governing America's Schools: The Shifting Playing Field." *Teachers College Record* 102/1.
Murphy, J., Beck, L. G., Crawford, M., Hodges, A., and McGaughy, C. L. (2001). *The Productive High School: Creating Personalized Academic Communities*. Thousand Oaks, CA: Corwin Press.
Mutua, K. and B. B. Swadener (2004). *Decolonizing Research in Cross-Cultural Contexts: Critical Personal Narratives*. Albany: State University of New York.
Mythen, G. (2004). *Ulrich Beck: A Critical Introduction to the Risk Society*. London: Pluto Press.
National Advisory Committee on Māori Education (NACME) (1980). *He Huarahi*. Wellington: Government Printer.
National Education Association (NEA). (2003). *Rural Education*. National Education Association. http://www.nea.org/rural. Accessed May 29, 2007.
National Indian Brotherhood (1972). *Indian Control of Indian Education*. Ottawa: National Indian Brotherhood.
National Research Council (NRC). (2002). "Studying Classroom Teaching as a Medium for Professional Development." In *Proceedings of a U.S.–Japan Workshop*. Ed. H. Bass, Z. Usiskin, and G. Burrill. Washington, DC: National Academy Press.
National Research Council and the Institute on Medicine. (2004). *Engaging Schools: Fostering High School Students' Motivation to Learn*. Washington, D.C: The National Academies Press.
National Science Board (NSB). (2006). *Science and Engineering Indicators 2006*. Washington, DC: National Science Foundation.
National Staff Development Council (NSDC). (2004). *Building for Success: State Challenge Grants for Leadership Development*. Report of a study. Seattle, WA: Bill and Melinda Gates Foundation.
Neill, A. S. (1937). *That Dreadful School*. London: Herbert Jenkins.
———. (1939). *The Problem Teacher*. London: Herbert Jenkins Limited.
———. (1968). *Summerhill*. Harmondsworth: Penguin (First published 1962).
———. (1971). *Talking of Summerhill*. London: Gollancz.
Nesbitt, E. and Henderson, A. (2003). "Religious Organisations in the U.K. and Values Education Programmes for School." *The Journal of Beliefs and Values* 24: 75–88.
Newell, R. (2005). "Student ownership: Teacher ownership." In *The coolest school in America: How small learning communities are changing everything*. Ed. D. Thomas, W. Enloe and R. Newell. Lanham, MD: Scarecrow Press.
Newmann, F. M. (1993). "Beyond Common Sense in Educational Reculturing: The Issues of Content and Linkage." [Electronic version] *Educational Researchers* 22/2: 4–13+22.

Newmann, F. M. (1996). *Authentic Achievement: Restructuring Schools for Intellectual Quality*. San Francisco: Jossey-Bass Publishers.

———. (2000). "Authentic Intellectual Work: What and Why?" *Research/Practice* 7/1. http://education.umn.edu/carei/Reports/Rpractice/Fall2000/default.html. Accessed March 16, 2004.

Newmann, F. M., Bryk, A. S., and Nagaoka, J. K. (2001). *Authentic Intellectual Work and Standardized Tests: Conflict or Coexistence?* Chicago: Consortium on Chicago School Research.

Newmann, F. M. and Wehlage, G. G. (1995). *Successful School Restructuring: A Report to the Public and Educators*. Madison, WI: University of Wisconsin, Center on Organization and Restructuring of Schools.

Ng, R. (1993). "Racism, Sexism and Nation Building in Canada." In *Race, Identity and Representation in Education*. Ed. C. McCarthy and W. Crichlow. New York: Routledge.

Nir, A. E. and Inbar, D. (2004). "From Egalitarianism to Competition: The Case of the Israeli Educational System." In *Balancing Change and Tradition in Global Education Reform*. Ed. I. Rotberg. Lanham, Maryland: Scarecrow.

Nock, D. (1988). *A Victorian Missionary and Canadian Indian Policy*. Waterloo: Wilfred Laurier Press.

Noddings, N. (1996). "On Community" [Electronic version]. *Educational Theory* 46/3: 245–267.

Nordwall, S. (1980). Rudolf Steiner and the Threefold Social Order. Text originally published in *Nordic Journal for Anthroposophical Medicine* 1. Available from author at http://www.thebee.se/indexeng.htm

Oberman, I. (1997). Waldorf History: Case Study of Institutional Memory. Paper presented at American Educational Research Association Annual Conference, Chicago.

OfSTED (2007). *Inspection Report: Dharma School, 12–13 June*. London: Office for Standards in Education.

Ogletree, E. (1998). *International Survey of the Status of Waldorf Schools*. (source: ERIC.(Educational Resources Information Center)). Illinois: University of Illinois.

———. (2000). "Creative Thinking Development of Waldorf School Students: A Study." *Trans-Intelligence Magazine* 7. Available on-line at http://www.transintelligence.org/articles/Creative%20Thinking%20Development%20of%20Waldorf%20School%20Students.htm.

O'Hair, M. J., McLaughlin, H. J., and Reitzug, U. C. (2000). *Foundations of Democratic Education*. Fort Worth, TX: Harcourt Brace.

O'Hair, M. J. and Reitzug, U. C. (2006). "Working for Social Justice in Rural Schools: A Model for Science Education." *International Electronic Journal for Leadership in Learning, 10*. Accessed December 18, 2006, http://www.ucalgary.ca/~iejll/volume10/o%27hair_reitzug.htm.

O'Hair, M. J., Reitzug, U. C., Cate, J., Averso, R., Atkinson, L., Gentry, D., Garn, G., and Jean-Marie, G. (2005). "Networking for Professional Learning Communities: School-University-Community Partnerships to Enhance Student Achievement." In *Network Learning for Educational Change*. Ed. W. Veugelers and M. J. O'Hair. London: Open University Press.

O'Hair, M. J. and Veugelers, W. (2005). "The Case for Network Learning." In *Network Learning for Educational Change*. Ed. W. Veugelers and M. J. O'Hair. Maidenhead: Open University Press.
Orange, C. A. (1987). *The Treaty of Waitangi*. Allen and Unwin, Port Nicholson: Wellington.
Oriaro, C. A. (2006). Spiritual, Moral and Cultural Values of Male Students in Two Faith-Based Schools: A Christian School and an Islamic School. Ph.D. dissertation. Biola University.
Osler, A. (2002). *Citizenship and the Challenge of Global Education*. Stoke-on-Trent: Trentham.
Osler, A. and Starkey, H. (2005). *Changing Citizenship. Democracy and Inclusion in Education*. Maidenhead: Open University Press.
Parekh, B. (2006). *Rethinking Multiculturalism*. Basingstoke: Palgrave Macmillan.
Pasley, J. D., Weiss, I. R., Shimkus, E., and Smith, P. S. (2004). "Looking Inside the Classroom: Science Teaching in the United States." *Science Educato* 13/1: 1–11.
Penn, W. (1726). *A Collection of the Works of William Penn*. Sowle: London.
Petrash, J. (2003). *Understanding Waldorf Education*. Edinburgh: Floris Books.
Pettigrew, T. F. (1998). "Intergroup Contact Theory." *Annual Review of Psychology* 49: 65–85. Pettigrew, T. F. and Tropp, L. R. (2000). "Does Intergroup Contact Reduce Prejudice? Recent Meta-analytic Findings." In *Reducing Prejudice and Discrimination*. Ed. S. Oskamp. Mahwah, NJ: Earlbaum.
Piaget, J. (1962). *Play, Dreams, and Imitation in Childhood*. New York: Norton.
———. (1972). *The Psychology of the Child*. New York: Basic Books.
Piper, H. (2002). *Seeking Guidelines on Touch*. Unpublished Research Report, Manchester: Manchester Metropolitan University.
Preston, M. (2006). "The Most 'Representative' State: Wisconsin." *Politics: The Morning Grind*. http://www.cnn.com/2006/POLITICS/07/27/mg.thu/index.html. Accessed August 22, 2006.
Pring, R. (2007). "The Common School." *Journal of Philosophy of Education* 41/4: 503–522.
Prosser, D. (2004). An Illuminative Evaluation of the Year-10 Bi-lingual Unit. A research project presented in partial fulfilment of the requirements for the degree of Master of Educational Administration at Massey University, Palmerston North, New Zealand.
Pullman, P. (2004). "The Art of Reading in Colour." *Index on Censorship* 33/4: 156–163.
Quaker Faith and Practice. (1994; revised edition, 2005). Quaker Faith and Practice: The Book of Christian Discipline of the Yearly Meeting of the Religious Society of Friends (Quakers) in Britain. London: Yearly Meeting 2004. Quaker Schools Web site www.quakerschools.co.uk/htm. Accessed May 1, 2007.
Ramsay, P. (1972). "Māori Schooling." In *Issues in New Zealand Education*. Ed. S. J. Havill and D. R. Mitchell. Auckland: Hodder and Stoughton.
Rathunde, K. (1995). "Wisdom and Abiding Interest: Interviews with Three Noted Historians in Later Life." *Journal of Adult Development* 2/3: 159–172.
———. (1996). "Family Context and Talented Adolescents' Optimal Experience in School-Related Activities." *Journal of Research on Adolescence* 6/4: 603–626.

Rathunde, K. (2001a). "Family Context and the Development of Undivided Interest: A Longitudinal Study of Family Support and Challenge and Adolescents' Quality of Experience." *Applied Developmental Science* 5: 158–171.

———. (2001b). "Montessori Education and Optimal Experience: A Framework for New Research." *NAMTA* 26/1: 11–43.

Rathunde, K. and Csikszetnmihalyi, M. (2005a). "Middle School Students' Motivation and Quality of Experience: A Comparison of Montessori and Traditional School Environments." *American Journal of Education* 111/3: 341–371.

———. (2005b). "The Social Context of Middle School: Teachers, Friends, and Activities in Montessori and Traditional School Environments." *Elementary School Journal* 106/1: 59–79.

———. (2006). "The Developing Person: An Experiential Perspective." In *Handbook of Child Psychology: Vol.1. Theoretical Models of Human Development*. Sixth edition. Ed. R.M. Lerner. Series Ed. W. Damon. New York: Wiley.

Ravenscroft, A. and McAlister, S. (2006). "Digital Games and Learning in Cyberspace: A Dialogical Approach." *E-Learning* 3/1: 37–50.

Rawson, M. and Richter, T. (2000). *The Educational Tasks and Content of the Steiner Waldorf Curriculum*. Forest Row, Sussex: Steiner Waldorf Schools Fellowship.

Reitzug, U. C. and O'Hair, M. J. (2002). "Tensions and Struggles in Moving toward a Democratic School Community." In *School as Community: From Promise to Practice*. Ed. G. Furman. Albany, NY: State University of New York Press.

Rivers, I. and Soutter, A. (1996). "Bullying and the Steiner School Ethos." *School Psychology International* 17: 359–377.

Rockquemore, K. A. and Schaffer, R. H. (2000). "Toward a Theory of Engagement: A Cognitive Mapping of Service-Learning Experiences." *Michigan Journal of Community Service Learning* 7: 14–25.

Roeser, R. W., Midgley, C., and Urdan, T. (1996). "Perceptions of the School Psychological Environment and Early Adolescents' Self-Appraisals and Academic Engagement: The Mediating Role of Goals and Belonging." *Journal of Educational Psychology* 88/3: 408–422.

Rofes, E. (2004). "Charter Schools as the Counterpublics of Disenfranchised Communities: Pedagogy of Resistance or False Consciousness?" In *The Emancipatory Promise of Charter Schools: Toward a Progressive Politics of School Choice*. Eds. E. Rofes and L. M. Stulberg. Albany: SUNY Press.

Rofes, E. and Stulberg, L. M. (Eds) (2004). *The Emancipatory Promise of Charter Schools: Toward a Progressive Politics of School Choice*. Albany: SUNY Press.

Rogoff, B. (1990). *Apprenticeship in Thinking : Cognitive Development in Social Context*. New York: Oxford University Press.

Rovai, A. P. (2002). "Building Sense of Community at a Distance." *International Review of Research in Open and Distance Learning* 3(1). http://www.irrodl.org/index.php/irrodl/article/view/79/152. Accessed August 1, 2006.

Royal Commission on Aboriginal peoples. (1996a). *Report of the Royal Commission on Aboriginal Peoples: Gathering Strength*. 3/120. Ottawa: Canada Communications Group.

———. (1996b). *Report of the Royal Commission on Aboriginal Peoples: Gathering Strength* 3/127. Ottawa: Canada Communications Group.

———. (1996c). *Report of the Royal Commission on Aboriginal Peoples: Gathering Strength*. 3/153. Ottawa: Canada Communications Group.
———. (1996d). *Report of the Royal Commission on Aboriginal Peoples: Gathering Strength*. 3/488. Ottawa: Canada Communications Group.
———. (1996e). *Report of the Royal Commission on Aboriginal Peoples: Perspectives and Realities*. 4/184. Ottawa: Canada Communications Group.
———. (1996f) *Report of the Royal Commission on Aboriginal Peoples: Renewal: A twenty-year commitment*. 5, Ottawa: Canada Communications Group.
Rusch, E. A. (1995). Leadership in Evolving Democratic School Communities. A paper presented at the annual meeting of the American Educational Research Association, San Francisco, CA. (ERIC Document Reproduction Service No. ED 392 117).
Sackney, L and Mitchell, C. (forthcoming). "Education in the 21ST Century: A New Paradigm for School Development." *Phi Delta Kappa*.
Sarroub, L. K. (2005). *All American Yemeni Girls: Being Muslim in a Public School*. Philadelphia: University of Pennsylvania Press.
Scheffler, I. (1984). "On the Education of Policymakers." *Harvard Educational Review* 54/2: 152–165.
Schein, E. (1992). *Organizational Culture and Leadership*. San Francisco: Jossey-Bass.
Schieffer, J. and Busse, R. (2001). "Low SES Minority Fourth-Graders' Achievement." *Research Bulletin* 6/1. Available at www.waldorflibrary.org/ResearchBulletin.htm
Schmidt, G. (2004). *Islam in Urban America: Sunni Muslims in Chicago*. Philadelphia: Temple University Press.
Schön, D. A. (1993). *The Reflective Practitioner – How Professionals Think in Action*. New York: Basic Books.
Secretaria Municipal de Educaçao de Porto Alegre (SMED). (2007). *Pesquisas e Informações Educacionais SMED/PMPA – Porto Alegre*. Available online at http://www2.portoalegre.rs.gov.br/smed.
Sergiovanni, T. J. (1994). *Building Community in Schools*. San Francisco: Jossey-Bass.
Simon, J. (1986). *Ideology in the Schooling of Māori Children. Delta Research Monograph, No. 7*. Palmerston North: Massey University.
———. (1994). "Historical Perspectives on Schooling." In *The Politics of Learning and Teaching in Aotearoa – New Zealand*. Ed. E. Coxon, K. Jenkins, L. Marshall, and L. Massey. Palmerston North: Dunmore Press.
Simon, J. and Smith, L. (2001). *A Civilising Mission? Perceptions and Representations of the New Zealand Native Schools System*. Auckland: Auckland University Press.
Sizer, T. R. (2004). *Horace's promise: The Dilemma of the American High School*. New York: Houghton Mifflin Company.
Skidelsky, R. (1969). *English Progressive schools*. Harmondsworth: Penguin.
Slater, J. and Cate, J. M. (2006). Cognitive Dissonance as a Perspective in the Transfer of Learning from Authentic Teacher Research Experiences to Inquiry Instruction in the Classroom. Paper presented at the American Educational Research Association Annual Conference, San Francisco, CA.
Smith, A., Lovatt, M., and Wise, D. (2005). *Accelerated Learning*. London: Crown House Publishing.

Smith, D. G. (2006). *Trying to Teach in a Season of Great Untruth: Globalization. Empire and the Crises of Pedagogy*. Rotterdam: Sense Publishers.

Smith, G. H. (1986). "Taha Maori: A Pakeha Privilege." *Delta* 37: 11–24 (June).

Smith, J. B., Lee, V. E. and Newmann, F. M. (2001). *Instruction and Achievement in Chicago Elementary Schools*. Chicago, IL: Consortium on Chicago School Research.

Smith, L. (1989). "Te Reo Maori: Maori Language and the Struggle to Survive." *Access* 8/1.

Smith, L. T. (1986). Seeing Through the Magic: Maori Strategies of Resistance. *Delta* 37: 3–10 (June).

———. (1999). *Decolonizing Methodologies: Research and Indigenous Peoples*. Dunedin, NZ: University of Otago Press.

Smith, P. (1998). "Update: Essentials of Waldorf Education Study." *Research Bulletin* 3/1. Available at www.waldorflibrary.org/ResearchBulletin.htm.

Smooha, S. (1996). "Ethno-Democracy: Israel as an Archetype." In *Zionism: A Contemporary Polemic*. Ed. P. Ginosar and A. Bareli. Jerusalem: BenGurion University (Hebrew).

Soder, R. (Ed.) (1996). *Democracy, Education and the Schools*. San Francisco: Jossey Bass.

Southwest Educational Development Laboratory (SEDL). (2004). *Technology Integration*. Austin, TX: Southwest Educational Development Laboratory.

Spiecker, B. and Steutel, J. (2001). "Multiculturalism, Pillarization and Liberal Civic Education in the Netherlands." *International Journal of Educational Research* 35: 293–304.

Spolsky, B. (1997). "Multilingualism in Israel." *Annual Review of Applied Linguistics* 17: 138–150.

Sprinzak, D., Segev, Y., Bar, E., and Levi-Mazloum, D. (2001). *Facts and Figures about Education in Israel*. Jerusalem: State of Israel, Ministry of Education, Economics and Budgeting Administration.

Stabiner, Karen (2002). *All Girls: Single Sex Education and Why It Matters*. New York: Riverhead Books.

Stake, R. E. (1978). "The Case-Study Method in Social Inquiry." *Educational Researcher* 7: 5–8.

Standing, E. M. (1984). *Maria Montessori: Her Life and Work*. New York: Penguin Books.

Starratt, R. J. (2003). *Centering Educational Administration: Cultivating Meaning, Community, Responsibility*. Mahwah, NJ: Lawrence Erlbaum Associates, Publishers.

Starratt, R. J. (2004). *Ethical Leadership*. San Francisco: Jossey-Bass.

Steele, C. M. (1997). "A Threat in the Air: How Stereotypes Shape Intellectual Identity and Performance." *American Psychologist* 52: 613–619.

Stewart, R. A. and Brendufer, J. L. (2005). "Fusing Lesson Study and Authentic Achievement." *Phi Delta Kappan* 86/9: 681.

Stewart-Harawira, M. (2005). *The New Imperial Order: Indigenous Responses to Globalization*. London: Zed Books.

Stokes, G. (2002). "Democracy and Citizenship." In *Democratic Theory Today*. Ed. A. Carter and G. Stokes. Cambridge: Polity Press.

Sumedho, A. (1992). *Cittaviveka: Teachings from the Silent Mind*. Great Gaddesden: Amaravati Publications.
Tam, H. (1998). *Communitarianism. A New Agenda for Politics and Citizenship*. London: Macmillan.
Tanesini, A. (2001). "In Search of Community. Mouffe, Wittgenstein and Cavell." *Radical Philosophy* 110: 12–19.
Tartter, V. C. (1996). *City College Report to FIPSE*. New York: City College Research Foundation.
Tatum, B. D. (1997). *"Why Are All the Black Kids Sitting Together in the Cafeteria?" And Other Conversations about Race*. New York, NY: Basic Books.
Taylor, Mark (Ed.). (1994). *Multiculturalism*. Princeton: Princeton University Press.
Tiger, L. (1999). "Hope Springs Eternal." *Social Research* 66/2: 611–623.
Tinirau, R. S. (2006). "Ngā Whakataetae mō Ngā Manu Kōrero: Ngā Manu Kōrero Speech Contest." In *He Wairere Pakihi: Māori Business Case Studies*. Ed. M.Mulholland. Te Au Rangahau/Māori Business Research Centre: Palmerston North.
Toch, T. (2003). *High schools on a Human Scale: How small schools can Transform American Education*. Boston: Beacon Press.
Trichur, Rita (2003). Islam and Toronto Public Schools: A Case of Contradictions in Canada's Multicultural Policy. M.A. thesis, Carleton University.
Turner, M. (2005). Maori in School Governance: The Voices of Māori Trustees. A research project presented in partial fulfilment of the requirements for the degree of Master of Educational Administration at Massey University, Palmerston North, New Zealand.
Uddin Sommieh (2004). "Hiring and Retaining Educators." *Islamic Horizons* 33/3 (May/June): 36–40.
UNDP (2007). *Human Development Report 2007–2008. Fighting Climate Change: Human Solidarity in a Divided World*. Available online at www.hdr.undp.org.
United States Department of Agriculture (USDA). Economic Research Service. (2004). "Rural Poverty at a Glance." *Rural Development Research Report Number 100*. http://www.ers.usda.gov/publications/rdrr100/rerr100.pdf. Accessed May 14, 2006.
Urion, C. (1999). Recording First Nations Traditional Knowledge. Unpublished paper, U'mista Cultural Society, 11. Cited in Stewart-Harawira (2005, 35).
Vaughan, M. (Ed.) (2006). *Summerhill and AS Neill*. London: Open University Press.
Vernon. A. (1982). *A Quaker Businessman: The Life of Joseph Rowntree 1836–1925*. London: William Sessions Ltd.
Vogelgesang, L. J. and Astin, A. W. (2000). "Comparing the Effects of Service-Learning and Community Service." *Michigan Journal of Community Service Learning* 7: 25–34.
Vygotsky, L. S. (1978). *Mind and Society: The Development of Higher Mental Processes*. Cambridge, MA: Harvard University Press.
Wagner, R. (1981). *The Invention of Culture*. Chicago: University of Chicago Press.
Wagtendonk, K. (1991). "Islamic Schools and Islamic Religious Education. A Comparison between Holland and Other West European Countries." In *The*

Integration of Islam and Hinduism in Western Europe. Ed. W. A. R. Shadid and P. S. van Koningsveld. Kampen: Kok.

Waitere-Ang, H. (1999). Te kete, The Briefcase, Te Tuara: The Balancing Act – Maori Women in the Primary Sector. Unpublished Master of Educational Administration thesis, Massey University.

Waitere-Ang, H. J. and Adams, P. J. (2005). "Ethnicity and Society." In *Education and Society in Aotearoa New Zealand*. 2. Ed. P. Adams, R. Openshaw, and J. Hamer. Nelson, VIC: Thompson/Dunmore Press.

Walker, R. (1985). "Cultural Domination of Taha Maori: The Potential for Radical Transformation." In *Political Issues in New Zealand Education*. Ed. J. Codd, R. Harker, and R. Nash. Palmerston North: Dunmore Press.

Walker, S. (1996). Kia Tau Te Rangi Marie: Kaupapa Maori Theory as a Resistance against the Construction of Maori as the Other. Unpublished M.A. thesis, Auckland University.

Walvin, J. (1997). *The Quakers: Money and Morals*. London, John Murray.

Ward, G. (2001). Education for the Human Journey: Personal Narrative in the Primary Classroom. Paper presented at Australian Association for Research in Education International Conference, Freemantle, 2–6 December.

Weber, M. (1978 [1956]). *Economy and Society*, Vols. I and II. Berkeley: University of California Press.

Wellman, B. (1999). "The Network Community: An Introduction to Networks in the Global Village." In *Networks in the Global Village*. Ed. B. Wellman. Boulder: Westview Press.

Wells, A. S. (1996). "Reexamining Social Science Research on School Desegregation: Long versus Short-Term Effects." In *Brown v. Board of Education: The Challenge for Today's Schools*. Ed. E. C. Lagemann and L. P. Miller. New York: Teachers College Press.

Wells, M. and Jones, R. (1998). "Relationship among Childhood Parentification, Splitting, and Dissociation: Preliminary Findings." *American Journal of Family Therapy* 26: 331–339.

Wentworth, R. (1999). *Montessori for the New Millennium*. Mahwah, NJ: Lawrence Erlbaum.

Werner, H. (1956). "On Physiognomic Perception." In *The New Landscape*. Ed. G. Kepes. Chicago: Theobald.

———. (1958). *Comparative Psychology of Mental Development*. New York: International Universities Press.

West, J. (1962). *The Quaker Reader*. Pendle Hill, PA: Pendle Hill Publications.

Westheimer, J. and Kahne, J. (2003). "Reconnecting Education to Democracy: Democratic Dialogues." *Phi Delta Kappan*, September, 9–14.

Wiliam, D. and Black, P. (2006). *Inside the Black Box: Raising Standards through Classroom Assessment*. London: NFER Nelson.

Williams, L. A. (2006). "The Influence of Technology Integration on High School Collaboration through the Development of a Professional Learning Community: A Mixed Methods Study." *Dissertations Abstract International* 67/04, AAT 3214722.

Williams, L., Atkinson, L., O'Hair, M. J., and Applegate, P. (2007, April). Improving Educational Quality through Technology-Enriched Learning

Communities for Success in the Global Economy. Paper presented at the meeting of the American Educational Research Association, Chicago, IL.
Williams, M. (2007). Feeling, Healing Transforming, Performing: Unsettling Emotions in Critical Transformative Pedagogy. Thesis presented as partial fulfilment of the requirements of the degree of Master of Education. Massey University.
Williams, R. (2005). *Culture and Materialism*. London: Verso.
Willinsky, J. (1998). *Learning to Divide the World: Education at Empire's End*. Minneapolis: University of Minnesota Press.
Wolf-Phillips, J. (2006). The Associative Leadership Toolkit. Unpublished paper.
Wood, G. H. (2005). *Time to Learn: How to Create High Schools That Serve All Students*. Second edition. Portsmouth, NH: Heinemann.
Woodhouse, J. and Knapp, C. (2000). "Place-Based Curriculum and Instruction: Outdoor and Environmental Education Approaches." *ERIC Digest*. Charleston, WV: ERIC Clearinghouse on Rural Education and Small Schools No. ED 448012.
Woods, G. J. (2003). Spirituality, Educational Policy and Leadership: A Study of Headteachers. PhD thesis. Milton Keynes: The Open University.
———. (2007). "The 'Bigger Feeling': The Importance of Spiritual Experience in Educational Leadership." *Educational Management Administration and Leadership*, 35/1: 135–155.
Woods, G. J., O'Neill, M., and Woods, P. A. (1997). "Spiritual Values in Education: Lessons from Steiner?" *International Journal of Children's Spirituality* 2/2: 25–40.
Woods, G. J. and Woods, P.A. (2008). "Democracy and Spiritual Awareness: Interconnections and Implications for Educational Leadership." *International Journal of Children's Spirituality* 13/2: 101–116.
Woods, G. J., Woods, P. A., and Ashley, M. (2005). *Building Bridges Conference: Summary of Outcomes – Towards a Wider Sense of Community?* Bristol: Faculty of Education, University of the West of England.
Woods, P. A. (2001). "Values-Intuitive Rational Action: The Dynamic Relationship of Instrumental Rationality and Values Insights as a Form of Social Action." *British Journal of Sociology* 52/4: 687–706.
———. (2004). "Democratic Leadership: Drawing Distinctions with Distributed Leadership." *International Journal of Leadership in Education: Theory and Practice* 7/1: 3–26.
———. (2005). *Democratic Leadership in Education*. London: Sage.
———. (2006). "A Democracy of All Learners: Ethical Rationality and the Affective Roots of Democratic Leadership." *School Leadership and Management* 26 /4: 321–337.
———. (2007a). "Authenticity in the Bureau-Enterprise Culture: The Struggle for Authentic Meaning." *Educational Management, Administration, and Leadership* 35/2: 297–322.
———. (2007b). "Within You and Without You: Leading Towards Democratic Communities."*Management in Education* 21/4: 38–43.
Woods, P. A. and Woods, G. J. (2002). "Policy on School Diversity: Taking an Existential Turn in the Pursuit of Valued Learning?" *British Journal of Educational Studies* 50/2: 254–278.
———. (2006a). *Feedback Report (Meadow Steiner School): Collegial Leadership in Action*. Aberdeen: University of Aberdeen.

Woods, P. A. and Woods, G. J. (2006b). *Feedback Report (Michael Hall Steiner School): Collegial Leadership in Action*. Aberdeen: University of Aberdeen.

———. (2006c). "In Harmony with the Child: The Steiner Teacher as Co-Leader in a Pedagogical Community." *Forum* 48/3: 217–325.

———. (2008). The Geography of Reflective Leadership: The Inner Life of Democratic Learning Communities. Paper presented at Philosophy of Management Conference, University of Oxford, July 11–14.

Woods, P. A., Ashley, M., and Woods, G. J. (2005). *Steiner Schools in England*. London: Department for Education and Skills. Ref No: RR645. Available at www.dfes.gov.uk/research.

Woods, P. A., Levacic, R., Evans, J., Castle, F., Glatter, R., and Cooper, D. (2007). *Diversity and Collaboration? Diversity Pathfinders Evaluation: Final Report*. London: Department for Education and Skills. Ref No: RR826. Available at www.dfes.gov.uk/research.

World Bank (2006). *World Development Indicators Database*. Available online at www.devdata.worldbank.org.

Yair, G. (2000). "Not Just about Time: Instructional Practices and Productive Time in School." *Educational Administration Quarterly* 36/4: 485–512.

York-Barr, J., Sommers, W. A., Ghere, G. S., and Montie, J. (2006). *Reflective Practice to Improve Schools: An Action Guide for Educators*, Second Edition. Thousand Oaks: Corwin Press.

Yoshida, M. (1999). Lesson study: A Case Study of a Japanese Approach to Improving Instruction through School-Based Teacher Development. Unpublished doctoral dissertation, University of Chicago.

Yount, D. (2007). *How the Quakers Invented America*. Boulder, Colorado: Rowman and Littlefield.

Zine, J. (2000). "Redefining Resistance: Towards an Islamic Subculture in Schools." *Race Ethnicity and Education* 3/3: 293–216.

———. (2004). Staying on the "Straight Path": A Critical Ethnography of Islamic Schooling in Ontario, PhD dissertation, University of Toronto.

Index

Aboriginal education, 167–85
Academic Performance Index (API), 24
accountability, 57, 60, 61, 64, 95, 104, 120, 145, 155, 157, 212, 224
acculturation, 137
achievement
  academic, 109, 119–20
  authentic, 29
  educational, 152
  goal(s), 205, 228
  school, 108–9, 112–13, 127
  student, 17, **18**, 22, 24, 27–8, **29**, 30, 48, 151, 178–9
  under-, 151, 152
  see also performance
administrators, 17, 74, 90, 109, 110, 120, 127, 151
aesthetic
  encounters, 193
  materials, 201
  phase of education, 211
  sense, 211
  sensibilities, 5
  sensitivities, 13, 232
  value, 21
affective
  capacities, 208
  dimensions/elements, 196, 239
  engagement, 207
  intensity, 208
  involvement, 199
  issues, 137
  mode, 199
  qualities, 207
affective-cognitive modes, 206
affirmative action, 129, 136
activist, 228, 229, **231**, 245, **246**
  'alter-native', 11, 144–5, 163
  'alternative', 11, 144–5, 153, 155, 157, 159, 163
  '(alter)native', 11, 145, 150, 153, 155, 157, 159, 163, 230, 234
  alternative, meanings of, 3–4, 7, 11, 83, 141, 143–4, 162, 227, 228–31, **231**
  engagement, 228, 229, **231**, 246, **246**
  orientations to societal context, 228–31, **231**
  separation, 228–9, **231**, 245, 246
anthroposophy, 12, 209–10, 219, 220
Arab, 122, 126
  see also Palestinian
Arabic (language), 10, 103, 125, 126, 128–9, 134
art, 13, 76, 83, 89, 103, 112, 197, 207, 210, 217–18, 224
assessment, 20, 21, 23, **29**, 50, 142, 212, 224
  standard assessment tests, 76, 222
Authentic Research Experiences (ARE), 23, **29**

Authenticity, 17, 26, 237
  authentic achievement, 29
  authentic assessment, 21
  authentic experiences, 18
  authentic intellectual work, 22–3
  authentic interactive instruction, 21–3, 24
  authentic learning, 18, **19**, **20**, 21, 22, 23
  authentic research, 18
  see also Authentic Research Experiences
authority, 6, 10, 37, 66, 86, 114, 116, 119, 211–12, 221, 241, 242, 247
autonomy, 59, 64, 84, 85, 97, 109, 110, 224, 238
  see also freedom

Batchelor, Stephen, 87–8
Battiste, Marie, 167, 171–7, 184, 185
behavior (students'), 52, 78, 95, 105, 134, 211
belief (religious), 83–4
benign panopticon, 59–60, 62
bhavana, 91
bhikkhus (monks), 86, 87
biculturalism, 142, 153
  Māori/Pākeha relations, 145, 148
  Treaty partners, 142, 162
  Treaty partnership, 163
  see also Treaty of Waitangi
bilingual
  additive bilingual approach, 126
  bilingualism, 128, 129, 136
  education, 123, 129, 135, 138, 143, 164
  educators, 129
  programs, 127, 130, 143, 144, 161
  schools, 10, 123, 127, 131, 132
  student (celebration of), 158
  teachers, 128, 230
Brock University, 179
Brown v. Board of Education, 136
Buddha, the, 87, 91, 92

bullying, 57, 61, 70, 78, 105, 215–16
bureau-enterprise culture, 235, **236**, 236, 238, 243, 245

Carey, Peter, 9, 86, 88–90
ceremony/ies, 66, 131, 132, 157, 160, 165, 171, 174, 180, 182, 186, 216, 244
Chah, Luang Por, 87
charisma, 96, 97
charismatic leadership, 97, 99
child democracy, 49
child development, 2, 49, 191, 199, 210–11, 213–14, 218, 219
choice, 2, 22, 32, 58, 77, 102, 128, 129, 130, 144, 181, 194, 198, 210, 221, 228, 237, 238
Christmas, 130
circlework, 11, 168, 169–70
class guardian, 211, 232
class teacher (Steiner/Waldorf schools), 211, 215, 223, 239–40
cognitive
  capabilities, 239
  development, 2, 91, 191, 227, 235, 236
  dimensions/elements, 196, 236, **246**
  engagement, 207
  faculties, 239
  involvement, 199
  learning, 190
  maturation, 12, 238
  psychology, 200
  skills, 13, **29**, 230, 232
  structures, 22
  see also serious/cognitive modes
collegiate system (Steiner/Waldorf), 211, 216, 221–2, 241, 244
colonization, 11, 143, 146, 167, 174, 183, 184, 186
  fourth world, 139, 143, 163, 165
compassion, 9, 86, 87, 88, 98, 99, 238
computers, 191, 211, 212

concentration, 89, 193–6, 199, 202, 205, 206, 238
conflict, 9, 58, 60, 71, 96, 109, 115, 123–4, 126, 130, 137, 177, 184, 219, 221, 247
conflict resolution, 9, 78, 93
connectedness, 13, 227, **234**, 239, 246–7 *see also* spiritual connectedness
Council of Ministers of Education of Canada, 178
creativity, 20, 76, 191, 192, 193, 195, 206, 207, 208
cross-cultural issues, 163
cultural capital, 108, 129, 152
curriculum integration, 13, 223, 238, 246

Dawkins, Richard, 85
decolonization, 101, 172, 181
  decolonizing methodologies, 163
  decolonizing work, 184–5
democracy, 5, 6, 16, 21, 22, 36, 37, 49, 62, 67, 75, 80, 84, 116, 124, 126, 222–3, 230, 231, 236, 237, 239
  developmental democracy, 4–5, 6, 234–46
  see also child democracy
democratic education, 4, 7, 8, 16, 27, 116, 235, **236**
democratic learning communities, 2, 3, 7, 16, 17, 18, 19, **19**, 21, 24, 26, 27, 28, 229, 237, 241, 242
democratic rationalities, 241–2, 244, 246
dentition (second), 211
Dewey, John, 21, 50, 61–2, 63, 72, 196, 199, 204, 207
Dhamma (in Pali ; in Sanskrit : Dharma) (truth or teaching), 87, 99
Dickason, Olive, 169
digital game-based learning, 19, 23, **29**
disembodied education, 12, 190–3

disembodiment and language, 190–2
drudgery, 12, 189, 192, 199, 203, 204, 205
dukkha (suffering), 87, 89, 91

economic sustainability, 99
educational leaders, 5, 25, 182, 240
embodied
  ceremony, 160
  cognition, 193, 200
  educational approaches, 193
  interactions, 60
  learning, 239–40, 243
  mind, 189, 191, 199, 200, 206
emotional, 181
  connection, 202, 213
  curriculum, 58
  development, 2, 91, 227, 235
  dimensions, 236, **246**
  engagement, 192
  ethos, 57
  experience, 239
  growth, 202
  intelligence, 243
  investment, 199, 205
  maturation, 12, 238
  reality, 12, 170, **170**, 184
  senses, 207
  synchrony, 191
emplaced education, 239–40, 242, 243, **246**
English language, 128, 129, 139, 141, 144, 149, 151, 152, 153, 154, 155, 158, 159, 216, 218
equity, 7, 16, 17, 18, 26, 138, 237
Ermine, Willie, 171
ethical
  development, 2, 227, 235
  dimensions, 12, 174, 236, **246**
  features of the environment, 51
  growth, 4
  rationality, 241, 242, **246**
  vision, 98

ethnicity, ethnic affiliation/group, etc., 22, 26, 61, 103, 104, 113, 125, 131, 132, 133–5, 153, 163, 221, 230
ethos, school, 1, 3, 57, 76, 77, 97, 98, 102, 103, 106, 113–15, 137, 216, 217, 235, 237
eurythmy, 218, 224
Experience Sampling Method (ESM), 196, 203

faith, 9, 67, 68, 79, 84, 85, 86, 90, 97, 98, 99, 104, 107, 108, 114, 115, 116, 117, 120, 146, 219
faith schools, 3, 4, 8, 9, 74, 83, 84–6, 98
feeling (stage in Steiner/Waldorf education), 211
feelings, 92, 132, 193–4, 198, 203–4, 211, 218, 238–9, 244
five precepts, 88, 92–3
flow, 12, 189, 193–9, 202, 203, 204, 205, 206, 238, 243, 246
fooling, 199, 203
Forest Retreat Order, 86, 87, 88
Fox, George, 66–7, 70, 80
freedom, 5, 6, 8, 12–13, 64, 69, 71, 101, 102, 107, 117, 118, 122, 172, 209–11, 214, 221, 232, **236**, 236, 237–42, 246
  and discipline, 196, 198–9, 202, 206, 243
fundamentalist position/stance, etc., 67, 221, 222

gender equality, 102, 117–19, 120, 121
geometric-technical thinking, 207
geopolitical context
  education in, 142
  see also Treaty of Waitangi
Goethe, 218–19
grammar of empathy, 56
grammar teaching, 216
Graveline, Fyre Jean, 168

Guided participation, 198

Hampton, Eber, 167, 170–3, 185, 186
Hanukkah, 130
Hardy, Sir Alister, 240
healing, 11, 12, 169, 177, 180, 181, 187
  see also Learning and Healing pedagogy
Hebrew (language), 10, 125, 126, 128, 129, 134
holistic
  approach, 213
  development, 6, 12, 13, 72, 76, 189, 205, 215, 218, 243
  education, 206
  engagement, 202
  firm framing, **236**, 236–7, 241, **246**
  growth, 247
  nature of personhood, 86, 238
  view, 196
  vision, 97
Hussar, Bruno, 123

IDEALS, 17–19, **18**, **19**, **20**, 27, 237
Idel-Fitter, 130
identities, 5, 54, 102, 115, 119, 120, 132, 133, 135, 232, 233
identity (religious), 106–7
inclusion, 119, 134, 157, 220
inclusivity, 96–7
Indian Control of Indian Education, 176
indigenous education, 143, 161
  assimilation, 150, 160
  integration, 150–1, 176
  Kōhanga Reo (preschool language nests), 155, 159, 166
  Kura kaupapa (immersion primary schools), 144, 145, 155, 157, 159, 161, 165
Māori education, 139–64
missionary boarding schools, 148
Native schools, 145, 149, 161

Taha Māori, 153
Wānanga (post compulsory higher education), 158, 166
  see also biculturalism, colonization, Māori/Pākeha, Ngā Manu Kōrero (speech competitions)
indigenous knowledge, 167, 177, 183, 184, 185
indigenous thought, 167–85
inner work, 211, 244
instrumental rationality, 6, 207, 228, 235
integration and differentiation, 196
integrative approach for freedom, 236, 237–42, 246
intrinsic cooperation, 233–4, 236, 245–6, 246
intrinsic motivation, 194, 195, 197, 198, 202, 203, 204, 205, 239
intuition, 8, 58, 192, 197, 207, 236, 239, 243
Islam, 10, 101, 103, 108, 114, 115, 116, 118, 120, 121, 242
Islamic ethos, 102, 106, 113–15
Islamic school boards, 108, 109, 111
Islamic schools, 9–10, 101–22
Israel, 10, 123, 124, 126, 127, 129, 131, 132, 137

Janus, 125
Jewish
  child(ren), 85, 129, 132, 134
  community/ies, 10, 124
  education sectors, 126
  hegemonic power in Israel, 131
  identity, 130
  liberal tradition, 124
  parents, 10, 125, 128, 129, 130, 131, 132
  population, 124, 128, 130
  staff, 10
  teachers, 126, 128, 130, 133
  traditions, 130

K20 Model, 7, 15–28, **18, 19, 20,** 29–30, 229, 237, 241
Kindergarten, 16, 124, 125, 133, 186, 211, 217

Land, 11, 131, 143, 148–9, 165, 168–9, 171, 172, 173, 176, 177, 180, 186, 187, 238, 239
Land Day, 132
leaders, 1, 5, **19,** 25, 26–7, 33, 96, 116, 148, 149, 207
  see also educational leaders
leadership, 5, 17, **18, 19, 20,** 24, 25, 26, 27, 88, 96, 98, 99, 107, 119, 152, 163, 224, 237
  see also charismatic leadership, educational leaders, leaders, school leaders
Learning and Healing pedagogy/theory, 12, 179, 180–1
lesson study, 18, **19,** 23, 29

main lesson, 13, 215, 223
mainstream curriculum, 157, 163
mandate system, 221, 224
Māori boarding schools, *see* indigenous education
Māori/Pākeha, 145, 148
Marples, Roger, 84–5
Massey University, 158–9
McCartney, Paul, 13
Mediation, 222
Medicine Wheel, 169–70, **170,** 172, 238
Memorial Day, 131
Mindful practice, **236,** 242–5, **246**
missionary boarding schools, *see* indigenous education
modern foreign languages, 216
Montessori, Maria, 189, 192, 193, 196, 199, 201, 205
multicultural
  awareness, 120
  contexts/society, 121, 142, 231, 233

multicultural—*continued*
  curriculum/education, 120, 124, 129, 135, 136, 137
  policy, 120, 121, 128
  posturing, 105
  strategies, 131
multicultural education, 120
multiculturalism, 4–6, 135, 142
  critical multiculturalism, 135
music, 76, 103, 112, 218
Muslim
  child(ren), 85, 105, 107, 108, 111, 113, 114, 116, 117, 120
  educators/teachers/staff, 103, 105, 108, 109, 122
  identity, 106, 107
  parents, 85, 105, 108, 109, 111, 115, 116, 119, 120, 229, 232
  village, 125
Muslims/Muslim population, 74, 101, 104, 105, 107–22, 130
mythology of Canada, 169

Naqbe, 131, 132
narrative, 59, 142, 146
nation-state, 125, 135, 136, 143, 165
National Curriculum
  England, 75, 91, 218, 223
  Netherlands, 108
national identity, 131, 132
National Indian Brotherhood, 176, 178
nature, 9, 201–2, 217, 218
networking, 19, 19, 20, 21, 23, 24, 25, 234
Ngā Manu Kōrero (speech competitions), 11, 139–41, 144, 148, 151, 155
noncognitive factors/dimensions, 113, 239
normalization, 196, 200, 206

observation
  meditative, 89, 243
  in Montessori education, 194–5, 197
  peer, 24

in Steiner/Waldorf education, 218
occult, 220
OfSTED (Office for Standards in Education), 49, 62, 64, 91, 95
Oklahoma, University of, 7, 16
oral culture, 147
oral tradition, 175, 216
oratory, 141, 144, 148, 155

Palestinian, 10, 123–35
panopticon, 52, 58, 64
pastoral support, 13, 223
peace, 9, 66, 70, 75, 78, 81, 103, 105, 124, 134, 195
  Great Law of Peace, 171
  Oasis of Peace, 123
pedagogical integration, 238, 246
pedagogical meetings, 212
Penn, William, 67
performance
  academic, 106, 108–10, 112–13, 121
  goal(s), 205
  measurable/measurement, 233, 235
  narrow, 241
  organizational, 243
  school, 21, 64, 76, 102, 104, 112–13
  tests, 206
phonics, 217
physiognomic perception, 207
Piaget, Jean, 22, 191, 196, 197, 219
PLANS, 220–1
play, 54, 56
Plessy v Ferguson, 136–7
prepared environment, 196, 239, 243
principals, 15, 16, 17, 19, 25, 32, 35, 36, 37, 38, 40, 41, 43, 45, 46, 47, 105, 106, 110–11, 113, 125
progressive education, 4, 63, 71, 72, 79, 85, 233, 235, 246
professional development, 1, 4, 15, 16, 17, 18, 19, 20, 21, 23–4, 25, 26, 27, 28, 29–30, 176, 178, 241
psychological complexity, 206–7, 243, 246

Quaker education
  controversial nature of schools, 73
  defining features, 74–8
  founding of schools, 71
  hope and resilience, 79
  modelling an alternative society, 78–80
  parental demands, 77
  school prospectuses, 65, 67, 76–8
Quakers
  American Quakerism, 69, 71, 80, 81
  common core of values, 68–9
  open and closed society, 70
  pacifism/peace testimony, 65, 68, 69, 75, 78
  protestants and the reformation, 66
  Quaker Faith and Practice, 67, 68, 71, 75, 76
  Quakers in Britain, 65, 67, 68, 69, 81
  social responsibility and action, 69–71
  worldwide, 69
quintessential spiritual experience, 7, 240, 243

Ramadan, 130
reflective activity(ies), 1, 13, 227, **231**, 234, 246
reflective practice, deepening, 1–3, 13, 227–47, 246
reflective teaching, 211
regeneration, 11, 168, **173**, 173, 175, 181–5
relational consciousness, 240, 243
relational touch, 57, 58, 60
relationships ('All My Relations', indigenous thought in Canadian education), 168–80
religion, 6, 13, 85, 87, 101, 107, 115, 117, 120, 122, 129, 210, 219, 247
remembering, 168, 172, **173**, 173, 174–6, 181, 238
Remembrance Day, 132

resistance, 8, 54, 146, 150, 152, 154, 168, **173**, 173, 176–7, 233, 238
respect, 5, 9, 12, 55, 66, 69, 73, 80, 94, 102, 104, 112, 115, 116, 120, 154, 164, 165, 171, 174, 178, 198, 199, 233, 247
responsibility, 22, **30**, 32, 53, 60, 65, 73, 168, 172, **173**, 173, 176, 177–80, 181, 211, 214, 220, 221, 238, 241, 245
risk culturalists, 56
risk culture, 57
Risk Society, 61
rituals, 54, 66, 211, 216–17, 244
Romans, 214
Rosh Hashanah, 130
routines, 50, 98, 103, 216–17, 241, 244
Royal Commission on Aboriginal Peoples, 178
rural schools, 7, 15, 16, 17, 25, 26

Scholastic Aptitude Test, 113
school leaders, 20, **20**, 25
science, 13, 15, 17, 23, 25, 103, 114, 155, 191, 192, 195, 206, 210, 218–19, 220
serious/cognitive modes, 199
service, 17, **18**, 26, 237
service learning, **19**, 19, 21, 24, **30**
shared vision, 19, 24, **30**, 237
skills, 2, 6, 13, **20**, 22, 26, 27, 28, **29**, **30**, 35, 56, 64, 73, 89, 113, 114, 115, 147, 150, 155, 156, 172, 186, 194, 195, 196, 197, 198, 212, 214, 216, 218, 230, 232, 238, 242, 246
skills and challenges, 195, 196–7, 199, 202, 206, 239, 243
social class, 221, 230
social interaction, 22, **29**, 132–3, 136
sociospiritual realities, 180
speech competitions, *see* Ngā Manu Kōrero

spirit, 6, 9, 66, 68, 69, 157, 166, 168, 172, **173**, 177, 186, 210, 228, 238, 244, 246
  spirit of the land, 168–70
  spirit-world, 180
spiritual
  affinities, 97
  attunement, 6
  awareness, 5, 66
  beings, 171
  beliefs, 180
  capability(ies), 7, 239
  connectedness, 240, 243, **246**
  connections, 11–12, 171
  development, 2, 6, 76, 89, 95, 97, 227, 235, 239, 242, 243
  dimensions, 12, 75, 88, 236, **246**
  endeavor, 114, 180
  environment, 181
  existence, 210
  experience, 240, *see also* quintessential spiritual experience
  growth, 4, 13, **30**, 215
  imperative, 170–80
  insight, 6, 180
  maturation, 12, 238
  place, 239
  prescription, 89
  principles, 237
  process, 12
  quality, 227
  reality, 12, **170**, 170, 171, 184, 244
  realms, 103
  reflection, 243
  sense, 66
  sensitivities, 13, 232
  sources, 217
  support, 104
  see also sociospiritual realities, spiritual-scientific system
spiritual, the, 236, 240, 243
spiritual-scientific system, 209

spirituality, 9, 13, 66, 74, 86, 147, 171, 179, 238
standard assessment tests (SATS), 76, 222
state funding, 101–2, 106, 109, 110, 112, 145, 157, 210, 221
State of Israel, *see* Israel
STEM education, 17, 25
Stewart-Harawira, Makere, 168, 174, 184, 186, 247
story, 143–4, 177, 205
storytelling, 174, 175, 211, 216
student achievement, *see* achievement
student engagement, 12, **19**, 20, 22, 189
subject knowledge, 212, 215
subject teacher, 215
Summerhillian
  Laws, 8, 52, 53, 58, 59, 60, 237
  the Meeting, 8, 50, 52–3, 55, 56, 57, 58, 59, 60, 237
  see also benign panopticon, child democracy, panopticon, relational touch, risk culturalists, risk culture, risk society, total institution, touching
supersensible knowledge, 219

talented adolescents, 207
technology-enriched learning community, 17, 20
technology integration, 21, 24
Ten Practices of High-Achieving Schools, 17, **18**, 19
testing, 120, 224
thinking (stage in Steiner/Waldorf education), 211
total institution, 52, 58
touching, 8, 50, 50–1, 57, 61, 63
transcendent
  awareness, 243
  power, 240

reality, 7, 239
values, 67, 70, 78
Treaty of Waitangi, 143, 146, 148, 156, 161, 165

undivided interest, 203, 204

value-rationality, 228
Vygotsky, Lev, 22, 59, 197

Werner, Heinz, 191, 192, 196, 206–7

willing (stage in Steiner/Waldorf education), 211
wisdom, 66, 87, 99, 199, 207, 209, 212, 222
Woods, Glenys, 4, 5, 6, 7, 12, 13, 210, 213, 215, 221, 222, 224, 233, 239, 240, 242, 244, 245

York University, 178, 186

zone of proximal development, 198